Long-Term Health State Estimation of Energy Storage Lithium-Ion Battery Packs

Qi Huang · Shunli Wang · Zonghai Chen ·
Ran Xiong · Carlos Fernandez · Daniel-I. Stroe

Long-Term Health State Estimation of Energy Storage Lithium-Ion Battery Packs

Springer

Qi Huang
Southwest University of Science and Technology
Mianyang, China

Shunli Wang
Southwest University of Science and Technology
Mianyang, China

Zonghai Chen
University of Science and Technology of China
Hefei, China

Ran Xiong
Southwest University of Science and Technology
Mianyang, China

Carlos Fernandez
Robert Gordon University
Aberdeen, UK

Daniel-I. Stroe
Aalborg University
Aalborg, Denmark

ISBN 978-981-99-5343-1 ISBN 978-981-99-5344-8 (eBook)
https://doi.org/10.1007/978-981-99-5344-8

This Springer imprint is published by the registered company Springer Nature Singapore Pte Ltd.
The registered company address is: 152 Beach Road, #21-01/04 Gateway East, Singapore 189721, Singapore

Preface

In recent years, the global application of new energy, particularly lithium-ion batteries, has experienced exponential growth due to the demands of the social economy. However, the internal structure of energy storage lithium batteries is highly complex, and their characteristics are strongly coupled, leading to the influence of various intricate factors such as temperature and degradation on the results of health state estimation. Achieving accurate and efficient long-term battery health state estimation is crucial to ensure the effective and stable utilization of new energy storage equipment across diverse operational conditions.

This book is crafted from the perspective of monitoring the long-term health state of lithium-ion batteries and aligns with the technical requirements of new energy storage power stations for energy storage lithium-ion batteries. It begins by addressing the electrochemical modeling of lithium-ion batteries, parameter identification of battery models, quantification of degradation modes, and estimation of long-term health state. Its primary focus lies in achieving accurate long-term health state estimation of lithium-ion batteries and packs through a combination of experiments, simulations, and data-driven algorithms. It extensively covers the key technologies involved in monitoring the health state of lithium batteries and packs, providing valuable technical references for the design and implementation of energy storage lithium battery management systems.

The book explores the establishment of battery pack electrochemical models and the identification of parameters. It introduces quantification methods for battery degradation modes, encompassing the quantification of active material loss and lithium inventory loss. Moreover, novel battery health state estimation methods for energy storage systems are developed, including long-term battery health state estimation and long-term battery pack health state estimation. The focus of this book is to address the long-term health state estimation challenges in the energy storage applications of lithium-ion batteries, making it an integral component of new energy and energy storage technology.

Based on electrochemical modeling, data-driven approaches, and the health state definition of lithium-ion batteries and battery packs, this book presents key technologies for long-term health state monitoring of lithium-ion batteries. It serves as a valuable technical reference for the design and implementation of energy storage lithium-ion battery management systems. With its distinctive features, systematic content, and ample examples, this book serves as an ideal textbook for disciplines such as control science, automation, electrical engineering, and other related fields in colleges and universities. Additionally, it serves as a valuable technical reference for researchers applying measurement and control technologies in the field of new energy sources.

The whole content is reviewed by Prof. Zonghai Chen, who provides lots of constructive opinions on the publication of this book. It has received scientific support from Mianyang quality supervision and inspection institute, Sichuan Huatai Electric Co., Ltd., Shenzhen Yakeyuan technology Co., Ltd., and Mianyang Weibo Electronics Co., Ltd.

As battery modeling involves a wide range of aspects, please feel free to contact the authors with the link https://www.researchgate.net/profile/Shunli-Wang-3 for effective responses. It is hoped that this book can be served as a communication platform to establish contact with readers and promote the progress of the core battery modeling and battery state estimation technology.

Mianyang, China Qi Huang
Mianyang, China Shunli Wang
Hefei, China Zonghai Chen
Mianyang, China Ran Xiong
Aberdeen, UK Carlos Fernandez
Aalborg, Denmark Daniel-I. Stroe

Acknowledgements

The work is supported by the National Natural Science Foundation of China (No. 62173281, 61801407), the Sichuan Science and Technology Program (No. 2019YFG0427), the China Scholarship Council (No. 201908515099), and the Fund of Robot Technology used for the Special Environment Key Laboratory of Sichuan Province (No. 18kftk03). Carlos Fernandez also expresses his profound gratitude to Robert Gordon University for its support. Thanks to these sponsors. Thanks are due to Shunli Wang for assistance with the experiments and to Ran Xiong for valuable discussion.

Contents

Chapter 1
Introduction

Abstract With the continuous optimization and advancement of internal materials, lithium-ion batteries have witnessed widespread application in various fields, including portable electronic devices, aerospace, new energy vehicles, and energy storage power stations. In particular, in the domain of energy storage, the rapid global development of new energy storage, led by lithium-ion batteries, has prompted countries worldwide to establish management regulations and guidance requirements, aiming to accelerate the progress of new energy storage technologies. This chapter provides a brief overview of the usage scenarios, research background, and research significance of energy storage lithium-ion batteries. Furthermore, it examines the current research status of lithium-ion battery cell electrochemical models (EMs), pack electrochemical models, and health state estimation methods by analyzing recent relevant references. Additionally, it introduces estimation methods for various state parameters of lithium-ion batteries, thereby laying a foundation for subsequent research endeavors. The system state estimation is emphasized to achieve the purpose of safety protection and health guarantee.

Keywords Lithium-ion battery · Energy storage · Battery model · Electrochemical model · Health state · System state estimation · Safety protection

1.1 Research Background

In recent years, there has been an exponential growth in the global adoption of new energy applications, particularly lithium-ion batteries, driven by socio-economic needs. The continuous optimization and enhancement of internal battery materials, coupled with significant advancements in their performance, have led to widespread utilization of lithium-ion batteries in various sectors, including portable electronic devices, aerospace, new energy vehicles, and energy storage power stations [1]. Within the realm of energy storage, lithium-ion batteries have gained immense popularity and are extensively employed in portable electronics, aerospace, new energy vehicles, and energy storage stations. The global development of new energy storage technologies, with lithium-ion batteries at the forefront, has been progressing rapidly.

Q. Huang et al., *Long-Term Health State Estimation of Energy Storage Lithium-Ion Battery Packs*, https://doi.org/10.1007/978-981-99-5344-8_1

However, when it comes to energy storage power stations based on lithium-ion batteries, there exists a lack of operational and maintenance expertise. This immaturity in the operational model has significantly impeded the growth and advancement of lithium-ion battery energy storage power stations.

Electrochemical energy storage technology holds a significant position within the realm of traditional energy storage technology [2, 3]. It has gained widespread popularity in new power systems due to its versatile siting options, environmentally friendly nature, and short construction cycles [4]. As power systems and energy markets undergo transformations to meet peak carbon and carbon–neutral objectives, electrochemical energy storage is set to become a crucial component of the power mix [5]. In particular, electrochemical energy storage technology has found extensive utilization and serves as the primary support in energy storage stations, especially in large-scale photovoltaic energy storage facilities [6–8]. Consequently, the application of electrochemical energy storage has emerged as a focal point and a challenging task in attaining the goals of peak carbon and carbon neutrality [9]. Thus, the application of electrochemical energy storage has become both the primary focus and a significant hurdle in achieving these objectives.

1.2 Research Significance

Electrochemical energy storage technology, represented by battery energy storage, has found extensive application in grid systems for large-scale energy storage. Lithium iron phosphate ($LiFePO_4$) batteries are preferred as the primary energy supply devices in new power systems due to their notable advantages of high stability, excellent performance, and resistance to high temperatures [10–16]. These batteries play a pivotal role in new power systems, providing various services such as new energy consumption, load compensation, and peak shaving, thereby delivering significant technical and societal benefits [17, 18]. As the core technology of new energy storage, Li-ion batteries offer diverse services, including new energy consumption, load compensation, and peak shaving. Consequently, the focus of electrochemical energy storage has gradually shifted towards market applications. However, with the rapid expansion of lithium battery energy storage operations, the operational reliability and safety of new energy storage stations have encountered significant challenges [19, 20]. Moreover, for new energy storage power stations, the safety and stability of the electric energy storage system components take precedence over the efficiency of energy supply. A substantial decline in the available capacity of any individual unit within the energy storage power plant can significantly impact overall performance [21]. Hence, it becomes crucial to establish an advanced battery management system (BMS) for monitoring the health state of the battery [22, 23]. A robust BMS serves as the foundation for ensuring the safety and reliability of new energy storage systems, and it holds important engineering value for the application of Li-ion batteries across various scenarios [24].

Accurate and efficient long-term health state estimation of batteries is essential to ensure the efficient and stable utilization of new energy storage devices under diverse working conditions. Health state serves as a metric for assessing battery degradation, diagnosing battery failures, and providing safety warnings. Technically, health state is a parameter that reflects the current health status of the battery, encompassing key performance indicators and enabling optimization of energy management strategies [25–27]. From an economic perspective, health state provides a reference for evaluating the economics of a battery, as the cost of battery wear is closely linked to the number of charge–discharge cycles it has undergone [28]. The definition of health state is primarily categorized into the internal resistance definition method and the capacity definition method [29–32]. Currently, various health state estimation methods have been proposed in the literature for energy storage lithium batteries. However, there is an urgent need to enhance these estimation methods and validate their applicability and effectiveness in practical engineering scenarios using current technologies [33]. Ensuring reasonable estimation of battery health state under stable operation within new power systems poses a significant challenge.

In conclusion, the health state of batteries holds great significance for new energy storage stations, and the accuracy and efficiency of the estimation results are vital for both system management strategies and the decision-making process of practitioners. Accurate health state estimation for energy storage lithium batteries also provides insights into the cost associated with battery degradation, contributing to the analysis of battery economics.

1.3 Research Status

1.3.1 Research Status of Lithium-Ion Battery Cell Model

Currently, both domestic and foreign researchers have proposed physical models for estimating the health state of lithium-ion batteries, which include equivalent circuit models and electrochemical models. Among these models, the equivalent circuit model is based on various circuit elements to simulate the dynamic voltage characteristics of the battery. Each parameter in the equivalent circuit model has a clear physical interpretation and can be expressed using simple mathematical equations [34]. In addition to the fundamental Theven in model and second-order RC model, researchers have proposed a series of enhanced composite models. For example, Zeng et al. [35] have developed an improved second-order RC model by incorporating high capacitance and a current-controlled current source. He et al. [36] have proposed a variable-parameter equivalent hysteresis model based on the Thevenin model and the hysteresis characteristics of the open-circuit voltage. Deng et al. [37] have further improved the first-order RC model by considering the current direction and hysteresis effects, building upon Baronti's study [38]. Liu et al. [39], Lai et al. [40], and Zhang et al. [41] have proposed fractional equivalent circuit models from different perspectives. Jin et al. [42] have analyzed the performance of both whole-order and fractional-order models in estimating battery state under different working conditions. Hu et al. [43] have compared the simulation performance of various equivalent circuit models under different working conditions and temperatures, and their findings suggest that the first-order RC model with hysteresis effect is highly suitable for describing lithium iron phosphate batteries. Therefore, this model deserves significant attention in the context of energy storage systems.

The EM, based on porous electrode theory and concentrated solution theory, provides an accurate simulation of the electrochemical reaction process occurring inside the battery cell. This approach allows the parameters of the electrochemical model to have clear physical interpretations. However, this type of model is characterized by high accuracy and complex structure. The pseudo-two-dimensional (P2D) model, which comprises several interconnected partial differential equations, considers the diffusion processes in the solid-phase, liquid-phase, and migration processes of the cell reaction [44]. The P2D model is commonly used as a reference mechanistic model for electrochemical modeling, laying the foundation for the advancement of EMs. The principles and challenges of conventional EMs are shown in Fig. 1.1.

The P2D model presents challenges due to its numerous nonlinear equations, making it computationally demanding and difficult to solve efficiently [45]. Moreover, the identification of a large number of parameters in the P2D model often leads to issues such as overfitting or local optimization problems. These challenges hinder its practical application in engineering, necessitating the need for simplification from an engineering perspective [46].

The simplification of electrochemical models can be achieved through two main methods. The first method involves reducing the order of the partial differential equation that describes the model, while the second method focuses on simplifying the internal structure of the model. For the first method, researchers have proposed

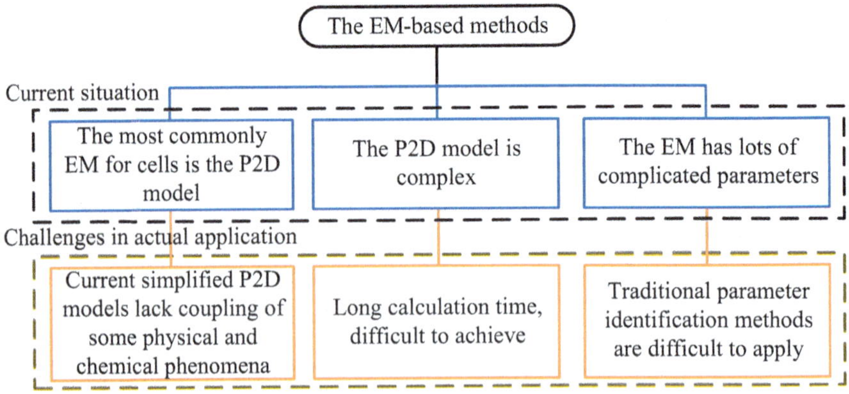

Fig. 1.1 The principles and challenges of conventional electrochemical models

various techniques. Reference [47] proposed an electrochemical model with a simplified side reaction transfer function. Lee et al. [48, 49] used a discrete-time implementation algorithm to reduce the five partial differential equations in the porous electrode theory of the P2D model. Deng et al. [50] applied the polynomial approximation principle to downscale the P2D model. The second method involves simplifying the internal structure of the model, and two models that have been proposed are the single particle (SP) model and the extended single particle (ESP) model [51]. The SP model is a simplified version of the P2D model, but it is less effective in simulating complex working conditions and high magnification. On the other hand, the ESP model builds upon the SP model by considering additional internal structure, reaction processes, and environmental factors, thereby improving the accuracy and generalization of the model. Several studies have further enhanced the SP and ESP models. For example, Mehta et al. [52] proposed an improved SP model that considered the spatial variation of overpotential and open circuit potential to estimate cell state. Lotfi et al. [53] introduced the electrolytic phase potential difference and kinetic factors into the SP model. Grandjean et al. [54] developed an electrolyte SP model that incorporated the effect of the electrolyte on the output voltage by introducing the conservation laws of liquid electrolyte substance and charge. Sadabadi et al. [55] investigated an ESP model that considered concentration overpotential to estimate the cell's health state. These simplified models provide alternative approaches that balance computational efficiency and model accuracy for practical engineering applications.

1.3.2 Research Status of Lithium-Ion Battery Pack Model

In the context of lithium-ion battery packs, the inconsistency between individual cells can significantly impact the overall performance. Modeling the battery pack is necessary because the behavior of the pack cannot be simply extrapolated from the behavior of individual cells. There are four main types of battery pack connections: series, parallel, series–parallel, and parallel-series, with the latter two referred to as mixed connections [56]. The principle of series–parallel connection indicates that series-connected battery packs increase the overall terminal voltage, while parallel-connected battery packs increase the overall capacity. In this context, battery packs refer to series-connected battery packs. When individual cells are connected in series, capacity inconsistency becomes a crucial factor affecting the energy and power of the battery pack, while internal resistance inconsistency affects the power of the battery pack. Various models have been proposed to represent battery packs, including the single-cell model (SCM), voltage maximum and voltage minimum model (VVM), and mean-difference model (MDM) model [30]. These models aim to capture the behavior and characteristics of the battery packs, taking into account factors such as capacity and internal resistance inconsistency. By using such models, the performance of the battery pack can be better understood and optimized.

The SCM can be further divided into the big-cell model (BCM) and multi-cell model (MCM) [30]. The BCM treats the battery pack as a single large cell and uses only the pack current and terminal voltage as input and output, without considering the internal structure and inter-unit inconsistency within the pack. Reference [57] has proposed a BCM that does not consider inconsistency. Another reference [58] equated a pack of cells with similar capacity and internal resistance to a first-order RC BCM. The BCM can capture the overall dynamic behavior of the battery pack but lacks the ability to characterize the internal characteristics of individual cells, which limits its adaptability to degradation effects. On the other hand, Matthieu et al. [59] developed an MCM that considers capacity and internal resistance differences among cells. The MCM takes into account cell inconsistency and provides highly accurate estimation results, suitable for describing the dynamic behavior of both individual cells and battery packs. However, the computational effort of the MCM increases with the number of cells, making it less suitable for online state estimation [30].

The VVM is a series battery pack model that selects the first overcharged cell during charging and the first overdischarged cell during discharging to characterize the state of the battery pack [30]. In reference [60], a relationship between the parameters of single cells and battery pack parameters are established using a series battery pack under different equalization control methods, aiming to construct a VVM that considers initial state of charge (SOC) differences, internal resistance differences, and battery capacity differences. Most of the single cell-based VVMs proposed in domestic and international literature are based on the overcharge and overdischarge characteristics of individual cells. However, there are also battery pack models that rely on a single characteristic for characterization. For instance, Hua et al. [61] proposed a specific battery pack VVM by considering the lowest available capacity and lowest voltage cell in the series battery pack as the lower limit. The VVM has a simple structure and can effectively prevent the battery pack from overcharging or over discharging. However, this model cannot describe the characteristics of each individual cell within the pack and may not accurately represent the overall state of the battery pack.

The MDM consists of a cell mean model (CMM) and several cell difference models (CDMs). The CMM characterizes the average behavior of the entire group, and the CDM characterizes the deviation between individual cells and the CMM. The CMM can be built based on electrochemical mechanisms or circuit principles. For example, reference [62] and reference [63] propose building the CMM using first-order equivalent circuit models and second-order circuit models, respectively. In existing literature, CDM includes five different types, with the complexity and accuracy of the model increasing as more parameters are considered. CDM can be based on an equivalent circuit model or neural networks. The first CDM considers only the difference in SOC between cells [30, 63, 64]. The second CDM considers both the difference in SOC and the difference in internal resistance [30, 64–66]. The third CDM incorporates the difference in available capacity between cells on top of the second CDM [30, 64]. The fourth CDM further considers the difference in resistance–capacitance (RC) parameters based on the third CDM [30, 64]. The fifth CDM, which is proposed in reference [30], considers the effects of twelve different parameters, resulting in high computational cost. Among these five CDMs, the second CDM is the most widely used cell difference model as it balances accuracy and computational effort well. The fourth and fifth CDMs have only been proposed theoretically and have not been further investigated by scholars. The MDM can effectively describe the dynamic behavior of individual cells as well as the overall group with low computational effort and high accuracy. However, as more inconsistency parameters are considered, the CDM becomes more complex, and using the CMM alone to describe the entire group may result in unsatisfactory accuracy. Therefore, future research on mean-difference models should focus on ways to characterize cell-to-cell inconsistency while ensuring computational simplicity.

1.3.3 Research Status of Health State Estimation Strategies

The health state of a single cell is often defined using the capacity definition method or the internal resistance definition method [67]. In the context of trams, the change in battery health state is primarily manifested as an increase in internal resistance, which reflects the degradation of energy capacity. In new energy storage systems, the change in battery health state is often characterized by capacity decay, where health state reflects the power capacity. Generally, when the capacity decay reaches 20% to 30%, the battery is considered to have reached the end of its life [68, 69]. There are also other proposed definitions of cell health state, such as monitoring the solid-phase diffusion time of recyclable lithium ions or lithium ions in the cathode to determine the battery's health state [70]. Regarding the calculation method of battery pack health state, Diao et al. [71] defined health state as the ratio of the maximum energy of the current battery pack to the initial energy. Building upon this definition, many international researchers have proposed various health state estimation methods from different perspectives, as illustrated in Fig. 1.2.

Figure 1.2 illustrates that the methods for estimating health state can be categorized into direct measurement methods and indirect analysis methods. The direct measurement methods primarily include the Coulomb counting method and the electrochemical impedance spectroscopy (EIS) method. The Coulomb counting method involves performing low-rate discharge experiments on the battery using charging and discharging equipment, and the measured data are used to calculate the health state after one cycle [72, 72]. However, this method is highly dependent on the

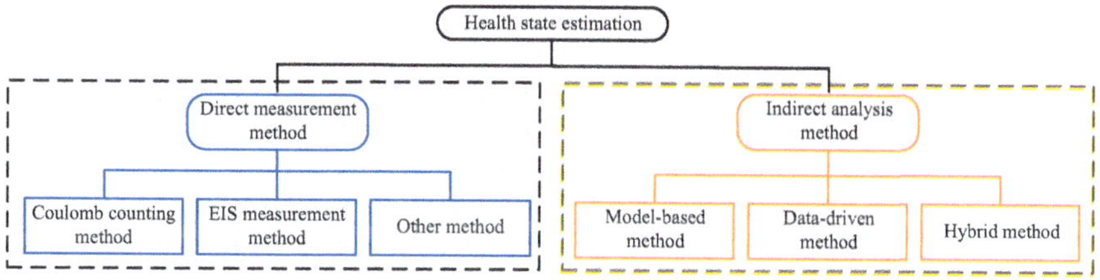

Fig. 1.2 Health state estimation methods from different perspectives

difference in SOC and current integration, leading to significant errors in the obtained health state. Moreover, error accumulation can occur over time [73]. The EIS method, on the other hand, utilizes the broad spectrum to assess health state. It calculates the electrochemical impedance, which is obtained from the ratio of the excitation voltage to the response current. Parameters such as charge transfer resistance and solid electrolyte interface (SEI) film resistance measured through EIS tests have been found to have the potential to estimate health state [74]. While the EIS method offers high-precision measurements and provides impedance-related information associated with electrochemical reaction processes, it requires specialized measurement techniques and extensive experimentation, making it challenging to apply online in engineering scenarios [75, 76].

In addition to the previously mentioned methods, there are other direct measurement methods for estimating health state of batteries. These include the internal resistance measurement method, cycle count method, and destructive test method. The internal resistance measurement method involves using specialized measuring instruments to obtain the battery's health state [77]. Since the internal resistance of a battery is typically very small, this method requires a precise measurement circuit to accurately determine the health state. The cycle count method determines the health state by counting the number of full charge and discharge cycles the battery has undergone [78]. However, this method has low estimation accuracy as it heavily relies on the rated number of discharges, and it may not be suitable for energy storage systems. The destructive test method involves disassembling the battery to study its degradation mechanisms and obtain the health state from a microscopic perspective [78]. Techniques such as X-ray diffraction [80] and X-ray photoelectron spectroscopy [81] have been used in these studies. However, these methods can cause irreversible damage to the battery, making them impractical for engineering applications. It is worth noting that while direct measurement methods can provide accurate health state estimation, they often require specialized equipment, careful experimental setup, and not be suitable for online or non-invasive monitoring in practical engineering scenarios.

The health state estimation method based on physical model can be divided from different perspectives, but all division methods cannot be separated from the internal reaction and external characteristics of the battery. The block diagram of health state estimation method based on physical model is shown in Fig. 1.3.

From Fig. 1.3, it can be observed that the process of health state estimation based on a physical model consists of three parts: model selection, parameter identification, and health state estimation. After appropriately modeling the lithium-ion battery using an equivalent circuit, the model parameters need to be identified, and then the health state is indirectly estimated using the identified parameters and selected state variables. Parameter identification for the equivalent circuit model can be divided into the curve fitting method and the least squares (LS) method [33]. The curve fitting method directly determines the direct current internal resistance by applying a pulse current and observing the resulting voltage variation. Additionally, the parameter values of polarization resistance and polarization capacitance can be obtained by comparing the time-domain circuit equations at the end of pulse charging and

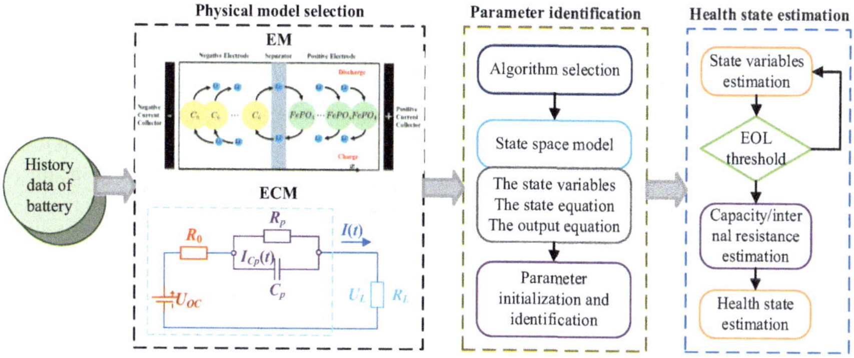

Fig. 1.3 The block diagram of health state estimation method based on physical model

discharging with the fitted curve. This method is an offline identification method, and although it is computationally simple, it may have larger errors. The least squares method continuously updates the model parameters to obtain the optimal solution based on the error between the model output and the actual output [79]. Among the various improvements to the LS method, recursive least squares (RLS) and its optimization methods are the most common online identification methods. In a study by Shi et al. [80], the RLS algorithm with a sliding window and differential forgetting factor was used for parameter identification. The results demonstrated that the method effectively utilizes new data for different operating conditions, and the identification process is computationally efficient with accurate results.

Once the model parameters are identified, the next step is to estimate the health state of the battery using a filtering algorithm. This involves constructing a state equation to model the state space and then iteratively solving for the health state dynamically. Two commonly used filtering algorithms for health state estimation are the extended Kalman filtering (EKF) [81] and particle filtering (PF) [82]. For example, Ma et al. [83] employed multi-innovative unscented Kalman and standard unscented Kalman methods to estimate the battery SOC and health state simultaneously. Based on the estimation of a fractional second-order RC model, they demonstrated the effectiveness of this cooperative estimation method. In another study, Bi et al. [84] developed a health state estimation method using a second-order RC battery pack equivalent circuit model and a genetic resampling particle filtering method. The usability of this model under typical on-board working conditions are verified. These filtering algorithms provide iterative solutions that improve the accuracy of health state estimation based on the identified model parameters. By leveraging the dynamics of the battery system and incorporating measurement data, these methods can effectively estimate the battery health state.

Similarly, in the electrochemical model-based approach, the identification of model parameters is a crucial step. A commonly used method for parameter identification is the multi-objective optimization algorithm. For instance, Wu et al. [72] employed an improved particle swarm optimization algorithm to identify two types of electrochemical parameters in the SP model. Moura et al. [85], on the other hand, developed an adaptive partial differential method observer to estimate the battery internal resistance and the number of recyclable lithium-ions. Estimating battery health state using electrochemical models often involves combining these models with filtering methods. The principles of filtering methods based on electrochemical models for health state estimation are similar to those based on equivalent circuit models. In these methods, the capacity or internal resistance is selected as the state variable. For example, Zheng et al. [86], Zou et al. [87], and Bartlett et al. [88] proposed different methods that combined proportional integral observer, multi-time scale observer, and a combination of EKF and PF, respectively, to estimate the resistance and capacity of the battery and obtain the health state of the lithium-ion battery. These studies verified the accuracy and robustness of their methods under various working conditions. By integrating electrochemical models with filtering methods, the battery health state can be estimated based on the identified parameters and the selected state variables, providing a comprehensive understanding of the battery's health condition.

Data-driven health state-based estimation methods can be classified from different perspectives [89, 90]. In this book, data-driven health state estimation methods based on machine learning (ML) can be further divided into two categories: ML-based methods and differential analysis-based (DA-based) methods. ML refers to the analysis of large datasets by automatically constructing models. The fundamental concept is to analyze the model through data learning and enable autonomous decision-making as much as possible. The general process framework of the ML-based health state estimation method is shown in Fig. 1.4.

Figure 1.4 depicts a general workflow for online health state estimation using ML techniques. The workflow comprises several key steps, including data acquisition, data preprocessing, estimation model development, and system integration analysis.

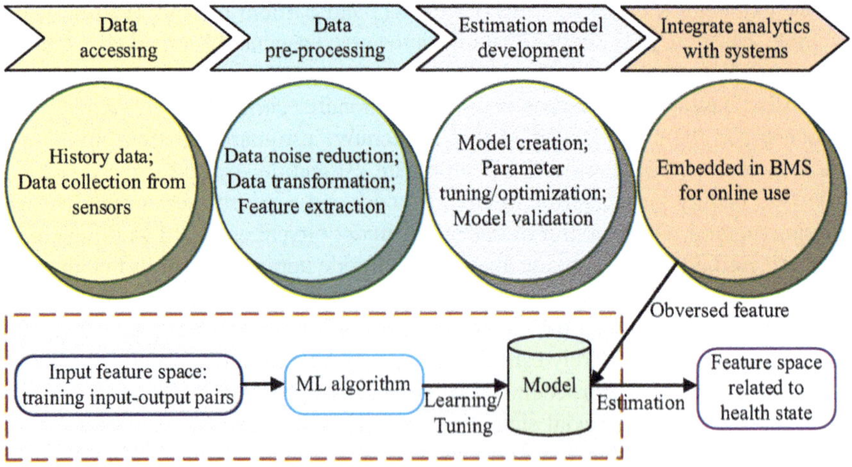

Fig. 1.4 The general process framework of the ML-based health state estimation method

ML methods commonly employed in health state estimation include artificial neural network (ANN), support vector machine (SVM), relevant vector machine (RVM), and fuzzy logic (FL).

ANN, as a Machine Learning method, is capable of simulating brain behavior and is often employed to model complex nonlinear systems such as batteries [91–93]. Within the realm of ANN, common types used for health state estimation include feedforward neural network (FFNN), Elman neural network (ENN), and long short-term memory (LSTM) neural network. Notably, both ENN and LSTM neural networks fall under the category of recursive neural network (RNN). In a study conducted by Pan et al. [94], the voltage at the constant current charging stage was fed into an FFNN model to estimate the health state of the battery. Another reference [95] utilized health indicators associated with lifetime degradation as inputs to an ENN model for health state estimation. Zhang et al. [96], after assessing the strengths and weaknesses of conventional RNN and LSTM structures, combined them to develop an adaptive time series prediction model for battery health state estimation.

SVM is a supervised ML method [97]. The fundamental concept behind SVM is to identify a small set of support vectors that can effectively represent the system from a large dataset. By utilizing a regression algorithm, SVM transforms a low-dimensional nonlinear model into a high-dimensional linear model [98]. In the domain of battery health state estimation, Nuhic et al. [99] presented an health state estimation method based on a SVM model that considers variations in temperature, SOC, and charge–discharge rate. Meng et al. [100] developed a two-layer structured SVM with radial basis functions, enabling efficient estimation of battery health state. Qin et al. [101] proposed an SVM model with enhanced degradation capture capability, employing a particle swarm-optimized kernel function for both remaining useful life (RUL) prediction and health state estimation.

RVM is a ML method based on the Bayesian framework theory. It achieves a sparse model through automatic relevance determination, which helps in rejecting irrelevant points. The resulting sparse model employs an arbitrary kernel function with high sparsity, enhancing computational efficiency. In the context of battery health state estimation, Wang et al. [102] utilized RVM to obtain a correlation vector capable of representing capacity degradation and estimating battery health state. Li et al. [103] developed a health state monitoring method that utilizes the optimal embedding dimension as the input for RVM. Qin et al. [104] proposed a method for both RUL prediction and health state monitoring using RVM, where the equal discharge voltage difference duration and equal charge voltage difference duration are used as inputs to the model.

FL is a bionic reasoning method that mimics the human brain and applies fuzzy logic theory to handle complex and nonlinear systems with fuzzy phenomena. It deals with problems by representing and reasoning with imprecise or uncertain information. In the realm of battery health state estimation, Pan et al. [105] proposed a fuzzy information granulation method for estimating the health state range. They used the time interval of equal charging current difference during constant voltage charging as input to the FL model. Landi et al. [106] employed an FL controller to fit the capacity decay curve and effectively estimate the health index at different working conditions. This approach indirectly represents the current health state of the battery.

DA-based methods focus on identifying characteristics related to capacity degradation by analyzing differential curves of electrical, thermal, or mechanical parameters during battery operation. These methods, such as incremental capacity analysis (ICA) method and differential voltage analysis (DVA) method, have gained significant attention in recent years due to their ability to study battery degradation mechanisms and estimate capacity loss without causing irreversible damage to the battery [107, 108]. ICA and DVA methods estimate health state by extracting features such as peaks, amplitudes, and envelope areas from the differentiation curves [109]. They offer a visual representation of battery health without the need for modeling. However, the differentiation curves may contain noise that cannot be completely eliminated, and the accuracy of DA-based methods heavily relies on low-rate constant current working conditions, which may not be representative of actual energy storage scenarios. To address some limitations, Naha et al. [110] proposed an online health state estimation technique for Li-ion batteries based on incremental voltage difference. This method overcomes the challenge of inaccurate health state estimation when charging data is limited by incorporating a feature vector that includes the difference between voltage and average temperature. Based on the DVA method, Berecibar et al. [111] presented a health state estimation method which is suitable for BMS, and validated the algorithm using a dataset of 18 individual cells.

Hybrid methods combine similar or different types of methods to estimate health state. In this book, hybrid models are classified into the method of combining direct measurement with physical model, the method of combining direct measurement with data-driven model, the method of combining data-driven with physical model, the method of combining different physical models, and the method of combining different data-driven models.

Regarding the method of combining direct measurement with physical model, Xiong et al. [112] developed a method that combines the EIS method and the dual polarization model for health state estimation of batteries. This hybrid approach leverages the advantages of impedance analysis and the physical insights provided by the dual polarization model. Regarding the method of combining direct measurement with data-driven model, Zenati et al. [113] introduced a hybrid method that combines the EIS method and FL to estimate the battery health state. This approach utilizes the measured impedance data and incorporates FL techniques to handle the complexity and uncertainty of battery behavior. Regarding the method of combining data-driven with physical model, Esfandyari et al. [114] and Park et al. [115] both proposed approaches that combine an equivalent circuit model with data-driven methods for health state estimation. Esfandyari et al. developed a joint health state and power state estimation method for a series-connected lithium-ion battery pack using model prediction and FL. Park et al. utilized the Thevenin model and a multivariate autoregressive algorithm to estimate the battery health state. Regarding the method of combining different physical models, Chu et al. [116] investigated a hybrid model that combines the electrochemical mechanism and the equivalent circuit model considering constant phase elements. This approach integrates different modeling perspectives to improve the accuracy of battery state estimation. Regarding the method of combining different data-driven models, Liu et al. [117] proposed a multi-feature fusion model based on the superposition algorithm, combining SVR and LSTM networks to estimate the battery health state. Similarly, Chen et al. [118] developed a hybrid model based on autoregressive sliding average and ENN for health state estimation. By summarizing and analyzing the above-mentioned types of lithium-ion battery health state estimation methods, Fig. 1.5 can be generated to illustrate the advantages, disadvantages, and potential improvement directions of different estimation methods.

Figure 1.5 provides valuable insights into the suitability and effectiveness of different battery health state estimation methods in various scenarios. Based on Fig. 1.5, several observations can be made:

(1) Direct measurement method, which involves directly measuring certain battery parameters, is found to be more suitable for laboratory scenarios. Direct measurements provide accurate and reliable information about the battery's health state but may require specialized equipment and controlled experimental setups.

(2) Physical model-based method, which utilizes mathematical models to describe battery behavior, is particularly useful in scenarios where there is limited data available, and a well-defined physical model can be constructed. This

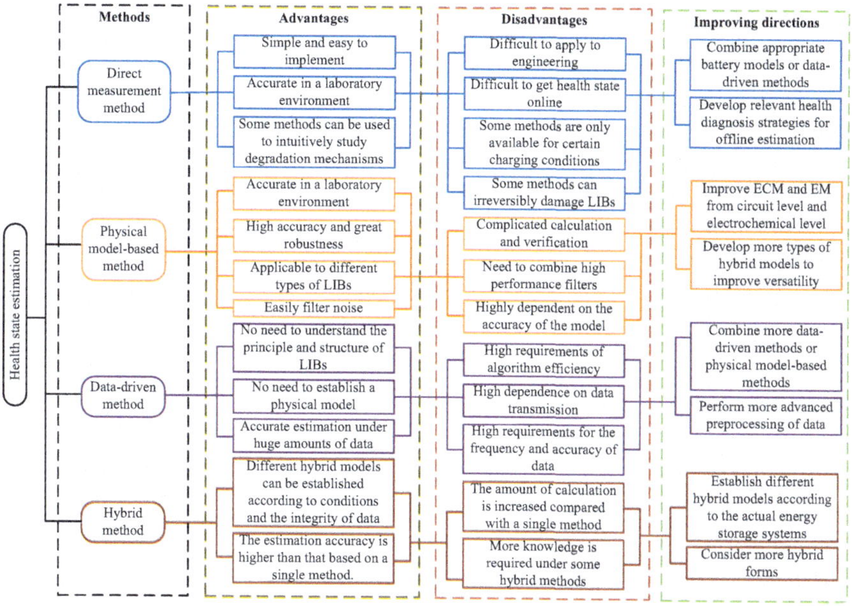

Fig. 1.5 Advantages, disadvantages, and improvement directions of different methods

method leverages the knowledge of battery physics and can provide insights into the underlying degradation mechanisms. However, it may require accurate knowledge of model parameters and assumptions about the battery's behavior.

(3) Data-driven method, which relies on machine learning and statistical techniques to analyze large datasets, is well-suited for engineering scenarios where ample data is available. By directly extracting health indicators from the collected data, this method can capture complex nonlinear relationships and adapt to various operating conditions. It offers flexibility and can handle diverse battery chemistries and usage patterns. However, it may require a significant amount of training data and can be sensitive to data quality and feature selection.

(4) Hybrid method, which integrates direct measurements, physical models, and data-driven techniques, emerges as a widely used and effective strategy in practical applications. By leveraging the strengths of multiple approaches, hybrid models can compensate for the limitations of individual methods and provide enhanced accuracy and robustness in estimating battery health state. This method offers a balance between accuracy and applicability across different scenarios.

Overall, Fig. 1.5 emphasizes the importance of selecting an appropriate estimation method based on the specific scenario, available data, and desired accuracy. It highlights the strengths, limitations, and suitability of different approaches and provides guidance for researchers and engineers in choosing the most appropriate method for battery health state estimation in different contexts.

1.4 Summary

This chapter provides an introduction to the research background and significance of the chosen topic in this book. Through an analysis of recent literature, it presents the current research status of the lithium-ion battery cell model, battery pack model, and health state estimation method. This sets the stage for readers to explore the intricacies of the battery modeling and the methodology behind estimating its health state in the forthcoming chapters.

References

1. Yang, N., et al. 2016. Unbalanced discharging and aging due to temperature differences among the cells in a lithium-ion battery pack with parallel combination. *Journal of Power Sources* 306 (29): 733–741.
2. Chen Xiaojiao, H.L., Liu Junbo, Song Dongran, and Yang Sheng. 2022. Peak shaving benefit assessment considering the joint operation of nuclear and battery energy storage power stations: Hainan case study. *Energy*, 239(PA): 1–15.
3. Ahmed Zayed, A.L Shaqsi, K.S., and Amr Al-Hinai. 2020. Review of energy storage services, applications, limitations, and benefits. *Energy Reports* 6(S7): 288–306.

4. Jaephil Cho, S.J., and Youngsik Kim. 2015.Commercial and research battery technologies for electrical energy storage applications. *Progress in Energy and Combustion Science* 48(6): 84-101.
5. Malcolm Abbott, B.C. 2020. Issues associated with the possible contribution of battery energy storage in ensuring a stable electricity system. *The Electricity Journal* 33 (6): 106771–106781.
6. Jin Yang, Z.Z., Miao Shan, Wang Qingsong, Sun Lei, and Lu Hongfei. 2021. Explosion hazards study of grid-scale lithium-ion battery energy storage station. *Journal of Energy Storage* 42(10): 102987-102998.
7. Olabi, A.G. 2017. Renewable energy and energy storage systems. *Energy* 136 (10): 1–6.
8. Bullich-Massagué, E., et al. 2020. A review of energy storage technologies for large scale photovoltaic power plants. *Applied Energy* 274 (15): 115213–115224.
9. Adua, L. 2020. Reviewing the complexity of energy behavior: Technologies, analytical traditions, and household energy consumption data in the United States. *Energy Research & Social Science* 59(C): 101289–101289.
10. Gholizadeh, M., and A. Yazdizadeh. 2020. Systematic mixed adaptive observer and EKF approach to estimate SOC and SOH of lithium–ion battery. *IET Electrical Systems in Transportation* 10 (2): 1–15.
11. Fang, L., J. Li, and B. Peng. 2019. Online estimation and error analysis of both SOC and SOH of Lithium-ion battery based on DEKF method. *Energy Procedia* 158 (1): 1–10.
12. Guo, Y., et al. 2021. A compact and optimized neural network approach for battery state of-charge estimation of energy storage system. *Energy* 219 (15): 119529–119541.
13. Hu, X., et al. 2017. Technological developments in batteries: a survey of principal roles, types, and management needs. *IEEE Power and Energy Magazine* 15 (5): 1–7.
14. Ran, X., et al. 2021. A novel joint estimation method of state of charge and state of health based on the strong tracking-dual adaptive extended Kalman filter algorithm for the electric vehicle Lithium-Ion batteries. *International Journal of Electrochemical Science* 16 (11): 1–32.
15. Yongli Wang, Y.S., Yuli Zhang, Xin Chen, Hekun Shen, Yang Liu, Xinya Zhang, and Yunxing Zhang. 2022. Optimal modeling and analysis of microgrid lithium iron phosphate battery energy storage system under different power supply states. *Journal of Power Sources* 521(15): 230931-230943.
16. Peng Qin, Z.J., Jingyun Wu, Kaiqiang Jin, Qiangling Duan, Lihua Jiang, Jinhua Sun, Jinghu Ding, Cheng Shi, and Qingsong Wang. 2022. The thermal runaway analysis on LiFePO4 electrical energy storage packs with different venting areas and void volumes. *Applied Energy* 313(1): 118767-118781.
17. Yushu Sun, W.P., Dongqiang Jia, Genming Zhang, Heng Wang, Long Zhao, and Zhizheng Feng. 2020. Application of integrated energy storage system in wind power fluctuation mitigation. *Journal of Energy Storage* 32(12): 101835-101843.
18. Akram, U., et al. 2020. A review on rapid responsive energy storage technologies for frequency regulation in modern power systems. *Renewable and Sustainable Energy Reviews* 120 (3): 109626–109638.
19. Jan, F., et al. 2021. The development of stationary battery storage systems in Germany – status 2020. *Journal of Energy Storage* 33 (9): 101982–101994.
20. Fatih, E., K. Mehmet, and V.A. Mete. 2021. Bidirectional DC–DC converter based multilevel battery storage systems for electric vehicle and large-scale grid applications: A critical review considering different topologies, state-of-charge balancing and future trends. *IET Renewable Power Generation* 15 (5): 915–938.
21. Wassiliadis, N., et al. 2018. Revisiting the dual extended Kalman filter for battery state-of-charge and state-of-health estimation: A use-case life cycle analysis. *Journal of Energy Storage* 19 (10): 73–87.
22. Liu, K., et al. 2019. Lithium-ion battery charging management considering economic costs of electrical energy loss and battery degradation. *Energy Conversion and Management* 195 (9): 167–179.
23. Xiaosong, H., Z. Kai, and L. Kailong. 2020. Advanced fault diagnosis for Lithium-ion battery systems: A review of fault mechanisms, fault features, and diagnosis procedures. *IEEE Industrial Electronics Magazine* 14 (3): 65–91.

24. Shunli, W., et al. 2022. A critical review of improved deep convolutional neural network for multi-timescale state prediction of Lithium-ion batteries. *Energies* 15 (14): 5053–5072.
25. Xu, L., Z. Deng, and X. Hu. 2021. Battery Health estimation using electrochemical aging model and ensemble Kalman filtering. *IEEE International Future Energy Electronics Conference* 1 (1): 1–6.
26. Xiaojun, T., et al. 2021. Online state-of-health estimation of lithium-ion battery based on dynamic parameter identification at multi timescale and support vector regression. *Journal of Power Sources* 484 (1): 1–10.
27. Sun, Y., et al. 2018. Remaining useful life prediction for lithium-ion batteries based on an integrated health indicator. *Microelectronics Reliability* 88–90 (9): 1189–1194.
28. Kang, B.O., et al. 2018. Economic analysis of a customer-installed energy storage system for both self-saving operation and demand response program participation in South Korea. *Renewable and Sustainable Energy Reviews* 94 (10): 69–83.
29. Fan, Y., et al. 2019. Experimental study on the thermal management performance of air cooling for high energy density cylindrical lithium-ion batteries. *Applied Thermal Engineering* 155 (5): 96–109.
30. Feng, F., et al. 2019. Propagation mechanisms and diagnosis of parameter inconsistency within Li-Ion battery packs. *Renewable and Sustainable Energy Reviews* 112 (1): 1–18.
31. Zhang, S., et al. 2020. A rapid online calculation method for state of health of lithium-ion battery based on coulomb counting method and differential voltage analysis. *Journal of Power Sources* 479 (15): 228740–228749.
32. Li, X., et al. 2019. State-of-health estimation for Li-ion batteries by combing the incremental capacity analysis method with grey relational analysis. *Journal of Power Sources* 410–411 (1): 106–114.
33. Tian, H., et al. 2020. A review of the state of health for lithium-ion batteries: Research status and suggestions. *Journal of Cleaner Production* 261 (10): 120813–120821.
34. Ren, H., et al. 2019. Design and implementation of a battery management system with active charge balance based on the SOC and SOH online estimation. *Energy* 166 (1): 908–917.
35. Zeng, M., et al. 2019. SOC and SOH joint estimation of the power batteries based on fuzzy unscented Kalman filtering algorithm. *Energies* 12 (16): 3122–3131.
36. He, Y., et al. 2021. Equivalent hysteresis model based SOC estimation with variable parameters considering temperature. *Journal of Power Electronics* 21 (3): 590–602.
37. Deng, Z., et al. 2016. Online available capacity prediction and state of charge estimation based on advanced data-driven algorithms for lithium iron phosphate battery. *Energy* 112 (1): 469–480.
38. Baronti, F., et al. 2015. Preisach modelling of lithium-iron-phosphate battery hysteresis. *Journal of Energy Storage* 4 (12): 51–61.
39. Changhe, L., et al. 2021. State of power estimation of lithium-ion battery based on fractional-order equivalent circuit model. *Journal of Energy Storage* 41 (9): 102954–102967.
40. Lai, X., et al. 2020. Co-estimation of state of charge and state of power for lithium-ion batteries based on fractional variable-order model. *Journal of Cleaner Production* 255 (10): 120203–120213.
41. Zhang, Q., et al. 2020. A novel fractional variable-order equivalent circuit model and parameter identification of electric vehicle Li-ion batteries. *ISA Transactions* 97 (2): 448–457.
42. Jin, G., et al. 2020. Comparison of SOC estimation between the integer-order model and fractional-order model under different operating conditions. *Energies* 13 (7): 1785–1797.
43. Hu, X., S. Li, and H. Peng. 2011. A comparative study of equivalent circuit models for Li-ion batteries. *Journal of Power Sources* 198 (15): 359–367.
44. Doyle, M., T.F. Fuller, and J. Newman. 1993. Modeling of Galvanostatic charge and discharge of the Lithium/Polymer/Insertion cell. *Journal of The Electrochemical Society* 140 (6): 1526–1539.
45. Shen, M., and Q. Gao. 2019. A review on battery management system from the modeling efforts to its multiapplication and integration. *International Journal of Energy Research* 43 (10): 5042–5075.

46. Wang, Y., et al. 2020. A comprehensive review of battery modeling and state estimation approaches for advanced battery management systems. *Renewable and Sustainable Energy Reviews* 131 (1): 110015–110024.
47. Zhang, X., et al. 2020. A novel quantitative electrochemical aging model considering side reactions for lithium-ion batteries. *Electrochimica Acta* 343 (20): 136070–136081.
48. Lee, J.L., A. Chemistruck, and G.L. Plett. 2012. One-dimensional physics-based reduced-order model of lithium-ion dynamics. *Journal of Power Sources* 220 (15): 430–448.
49. Lee, J.L., et al. 2014. Extended operating range for reduced-order model of lithium-ion cells. *Journal of Power Sources* 255 (1): 85–100.
50. Deng, Z., et al. 2018. Polynomial approximation pseudo-two-dimensional battery model for online application in embedded battery management system. *Energy* 142 (1): 838–850.
51. Lichao, R., et al. 2021. An algorithm for state of charge estimation based on a single-particle model. *Journal of Energy Storage* 39 (7): 102644–102651.
52. Rohit, M., and G. Amit. 2021. An improved single-particle model with electrolyte dynamics for high current applications of lithium-ion cells. *Electrochimica Acta* 389 (1): 138623–138631.
53. N. Lotfi, J.L., R. G. Landers and J. Park. 2017. Li-ion Battery State of Health Estimation based on an improved Single Particle model. *American Control Conference (ACC)* 1(1): 86–91.
54. T.R.B. Grandjean, L.L., M. X. Odio, and W. D. Widanage. 2019. Global Sensitivity Analysis of the Single Particle Lithium-Ion Battery Model with Electrolyte. *IEEE Vehicle Power and Propulsion Conference (VPPC)* 1(1): 1–7.
55. Sadabadi, K.K., X. Jin, and G. Rizzoni. 2021. Prediction of remaining useful life for a composite electrode lithium ion battery cell using an electrochemical model to estimate the state of health. *Journal of Power Sources* 481 (1): 228861–228871.
56. Cordoba-Arenas, A., S. Onori, and G. Rizzoni. 2015. A control-oriented lithium-ion battery pack model for plug-in hybrid electric vehicle cycle-life studies and system design with consideration of health management. *Journal of Power Sources* 279 (1): 1–13.
57. Sun, F., et al. 2012. Model-based dynamic multi-parameter method for peak power estimation of lithium–ion batteries. *Applied Energy* 96 (1): 1–15.
58. Xiong, R., et al. 2013. Adaptive state of charge estimator for lithium-ion cells series battery pack in electric vehicles. *Journal of Power Sources* 242 (1): 1–19.
59. Dubarry, M., N. Vuillaume, and B.Y. Liaw. 2008. From single cell model to battery pack simulation for Li-ion batteries. *Journal of Power Sources* 186 (2): 1–15.
60. Zhong, L., et al. 2014. A method for the estimation of the battery pack state of charge based on in-pack cells uniformity analysis. *Applied Energy* 113 (1): 1–15.
61. Hua, Y., et al. 2015. A multi time-scale state-of-charge and state-of-health estimation framework using nonlinear predictive filter for lithium-ion battery pack with passive balance control. *Journal of Power Sources* 280 (1): 1–16.
62. Li, J., et al. 2015. Multicell state estimation using variation based sequential Monte Carlo filter for automotive battery packs. *Journal of Power Sources* 277 (1): 1–15.
63. Dai, H., et al. 2012. Online cell SOC estimation of Li-ion battery packs using a dual time-scale Kalman filtering for EV applications. *Applied Energy* 95 (1): 1–15.
64. Feng F, H.X., and Liu K. 2020. A practical and comprehensive evaluation method for series-connected battery pack models. *IEEE Transactions on Transportation Electrification*, 6(2): 391–416.
65. Zheng, Y., et al. 2013. Cell state-of-charge inconsistency estimation for LiFePO4 battery pack in hybrid electric vehicles using mean-difference model. *Applied Energy* 111 (1): 1–15.
66. Zheng, Y., et al. 2018. State-of-charge inconsistency estimation of lithium-ion battery pack using mean-difference model and extended Kalman filter. *Journal of Power Sources* 383 (1): 1–15.
67. Sijia, Y., et al. 2021. Review on state-of-health of lithium-ion batteries: Characterizations, estimations and applications. *Journal of Cleaner Production* 314 (10): 128015–128024.
68. Xuebing, H., et al. 2019. A review on the key issues of the lithium ion battery degradation among the whole life cycle. *eTransportation*, 1(8): 100005–100014.

69. Chen, D., et al. 2020. Pomegranate-like Silicon-based anodes self-assembled by hollow-structured Si/void@C nanoparticles for Li-ion batteries with high performances. *Nanotechnology* 32 (9): 1–9.
70. Prasad, G.K., and C.D. Rahn. 2013. Model based identification of aging parameters in lithium ion batteries. *Journal of Power Sources* 232 (15): 79–85.
71. Diao, W., et al. 2017. Energy state of health estimation for battery packs based on the degradation and inconsistency. *Energy Procedia* 142 (11): 3578–3583.
72. Zhuoyan Wu, L.Y., Ran Xiong, Shunli Wang, Wei Xiao, Yi Liu, Jun Jia, and Yanchao Liu. 2022. A novel state of health estimation of lithium-ion battery energy storage system based on linear decreasing weight-particle swarm optimization algorithm and incremental capacity-differential voltage method. *International Journal of Electrochemical Science*, 17(7): 1-32.
73. Lashway, C.R., and O.A. Mohammed. 2016. Adaptive battery management and parameter estimation through physics-based modeling and experimental verification. *IEEE Transactions on Transportation Electrification* 2 (4): 454–464.
74. ShunLi, W., et al. 2021. An improved coulomb counting method based on dual open-circuit voltage and real-time evaluation of battery dischargeable capacity considering temperature and battery aging. *International Journal of Energy Research* 45 (12): 17609–17621.
75. Li, S.E., et al. 2014. An electrochemistry-based impedance model for lithium-ion batteries. *Journal of Power Sources* 258 (1): 9–18.
76. Westerhoff, U., et al. 2016. Electrochemical impedance spectroscopy based estimation of the state of charge of lithium-ion batteries. *Journal of Energy Storage* 8 (9): 244–256.
77. Yoon, S., et al. 2011. Power capability analysis in lithium ion batteries using electrochemical impedance spectroscopy. *Journal of Electroanalytical Chemistry* 655 (1): 32–38.
78. Mühlbauer, M.J., et al. 2018. Probing chemical heterogeneity of Li-ion batteries by in operando high energy X-ray diffraction radiography. *Journal of Power Sources* 403 (1): 49–55.
79. Xiong, R., L. Li, and J. Tian. 2018. Towards a smarter battery management system: A critical review on battery state of health monitoring methods. *Journal of Power Sources* 405 (30): 18–29.
80. Mühlbauer, M.J., et al. 2018. Probing chemical heterogeneity of Li-ion batteries by in operando high energy X-ray diffraction radiography. *Journal of Power Sources* 403 (1): 49–55.
81. Mühlbauer, M.J., et al. 2018. Probing chemical heterogeneity of Li-ion batteries by in operando high energy X-ray diffraction radiography. *Journal of Power Sources* 403 (1): 49–55.
82. Tsuda, T., et al. 2018. In situ electron microscopy and X-ray photoelectron spectroscopy for high capacity anodes in next-generation ionic liquid-based Li batteries. *Electrochimica Acta* 279 (1): 1–10.
83. Feng, G.M., et al. 2021. A review on state of health estimations and remaining useful life prognostics of lithium-ion batteries. *Measurement* 174 (4): 109057–109064.
84. Jinjin, S., G. Haisheng, and C. Dewang. 2021. Parameter identification method for lithium-ion batteries based on recursive least square with sliding window difference forgetting factor. *Journal of Energy Storage* 44 (15): 103485–103494.
85. Li, X., Z. Wang, and L. Zhang. 2019. Co-estimation of capacity and state-of-charge for lithium-ion batteries in electric vehicles. *Energy* 174 (1): 33–44.
86. Rui, X., et al. 2018. A double-scale, particle-filtering, energy state prediction algorithm for lithium-ion batteries. *IEEE Transactions on Industrial Electronics* 65 (2): 1526–1538.
87. Lili, M., et al. 2022. Co-estimation of state of charge and state of health for lithium-ion batteries based on fractional-order model with multi-innovations unscented Kalman filter method. *Journal of Energy Storage* 52 (15): 104904–104912.
88. Bi, J., et al. 2016. State-of-health estimation of lithium-ion battery packs in electric vehicles based on genetic resampling particle filter. *Applied Energy* 182 (15): 558–568.
89. Moura, S.J., M. Krstic, and N.A. Chaturvedi. 2012. Adaptive PDE Observer for Battery SOC/SOH Estimation. *ASME 2012 5th Annual Dynamic Systems and Control Conference joint with the JSME 2012 11th Motion and Vibration Conference*, 1(9): 101–110.
90. Zheng, L., et al. 2016. Co-estimation of state-of-charge, capacity and resistance for lithium-ion batteries based on a high-fidelity electrochemical model. *Applied Energy* 180 (15): 424–434.

91. Zou, C., et al. 2016. Multi-time-scale observer design for state-of-charge and state-of-health of a lithium-ion battery. *Journal of Power Sources* 335 (15): 121–130.
92. A. Bartlett, J.M., S. Onori, G. Rizzoni, X. G. Yang, and T. Miller. 2013. Model-based state of charge estimation and observability analysis of a composite electrode lithium-ion battery. *52nd IEEE Conference on Decision and Control*, 1(1): 7791–7796.
93. Li, Y., et al. 2019. Data-driven health estimation and lifetime prediction of lithium-ion batteries: A review. *Renewable and Sustainable Energy Reviews* 113 (10): 109254–109264.
94. Tsuyoshi, O., et al. 2021. Data-driven methods for battery SOH estimation: Survey and a critical analysis. *IEEE Access* 9 (1): 126903–126916.
95. Eddahech, A., et al. 2012. Behavior and state-of-health monitoring of Li-ion batteries using impedance spectroscopy and recurrent neural networks. *International Journal of Electrical Power and Energy Systems* 42 (1): 487–494.
96. Lipu, M.S.H., et al. 2018. A review of state of health and remaining useful life estimation methods for lithium-ion battery in electric vehicles: Challenges and recommendations. *Journal of Cleaner Production* 205 (20): 115–133.
97. You, G.-W., S. Park, and D. Oh. 2016. Real-time state-of-health estimation for electric vehicle batteries: A data-driven approach. *Applied Energy* 176 (15): 92–103.
98. Pan, H., et al. 2018. Novel battery state-of-health online estimation method using multiple health indicators and an extreme learning machine. *Energy* 160 (1): 466–477.
99. Zheng Chen, Q.X.Y.L., Jiangwei Sshen, and Renxin Xiao. 2019. State of health estimation for lithium-ion batteries based on elman neural network. *DEStech Transactions on Environment Energy and Earth Science*, 1(1): 1–10.
100. Zhang, W., X. Li, and X. Li. 2020. Deep learning-based prognostic approach for lithium-ion batteries with adaptive time-series prediction and on-line validation. *Measurement* 164 (11): 108052–108061.
101. Anton, A., et al. 2013. Support vector machines used to estimate the battery state of charge. *IEEE Transactions on Power Electronics* 28 (12): 5919–5926.
102. Hannan, M.A., et al. 2017. A review of lithium-ion battery state of charge estimation and management system in electric vehicle applications: Challenges and recommendations. *Renewable and Sustainable Energy Reviews* 78 (10): 834–854.
103. Nuhic, A., et al. 2013. Health diagnosis and remaining useful life prognostics of lithium-ion batteries using data-driven methods. *Journal of Power Sources* 239 (1): 680–688.
104. Meng, J., et al. 2018. Lithium-ion battery state of health estimation with short-term current pulse test and support vector machine. *Microelectronics Reliability* 88–90 (9): 1216–1220.
105. Qin, T., S. Zeng, and J. Guo. 2015. Robust prognostics for state of health estimation of lithium-ion batteries based on an improved PSO–SVR model. *Microelectronics Reliability* 55 (9–10): 1280–1284.
106. Wang, D., Q. Miao, and M. Pecht. 2013. Prognostics of lithium-ion batteries based on relevance vectors and a conditional three-parameter capacity degradation model. *Journal of Power Sources* 239 (1): 253–264.
107. Li, H., D. Pan, and C.L.P. Chen. 2014. Intelligent prognostics for battery health monitoring using the mean entropy and relevance vector machine. *IEEE Trans. Systems, Man, and Cybernetics: Systems,* 44(7): 851–862.
108. Qin, X., Qi Zhao, Hongbo Zhao, Wenquan Feng, and XiuMei Guan. 2017. Prognostics of remaining useful life for lithium-ion batteries based on a feature vector selection and relevance vector machine approach. *2017 IEEE International Conference on Prognostics and Health Management (ICPHM),* 1(1): 1–6.
109. Pan, W., et al. 2020. A data-driven fuzzy information granulation approach for battery state of health forecasting. *Journal of Power Sources* 475 (1): 228716–228724.
110. Landi, M., and G. Gross. 2014. Measurement techniques for online battery state of health estimation in vehicle-to-grid applications. *IEEE Trans. Instrumentation and Measurement* 63 (5): 1224–1234.
111. Yajun, Z., et al. 2022. State-of-health estimation for lithium-ion batteries by combining model-based incremental capacity analysis with support vector regression. *Energy* 239 (15): 121986–121995.

112. Dubarry, M., et al. 2011. Identifying battery aging mechanisms in large format Li ion cells. *Journal of Power Sources* 196 (7): 3420–3425.
113. M., M., et al. 2020. Lithium-ion battery SoH estimation based on incremental capacity peak tracking at several current levels for online application. *Microelectronics Reliability*, 114(11): 113798–113805.
114. Arunava, N., et al. 2020. An incremental voltage difference based technique for online state of health estimation of li-ion batteries. *Scientific Reports* 10 (1): 9526–9534.
115. Berecibar, M., et al. 2016. State of health estimation algorithm of LiFePO4 battery packs based on differential voltage curves for battery management system application. *Energy* 103 (15): 784–796.
116. Xiong, R., et al. 2017. A systematic model-based degradation behavior recognition and health monitoring method for lithium-ion batteries. *Applied Energy* 207 (1): 372–383.
117. A. Zenati, P.D.a.H.R. 2010. Estimation of the SOC and the SOH of li-ion batteries, by combining impedance measurements with the fuzzy logic inference. *IECON 2010 - 36th Annual Conference on IEEE Industrial Electronics Society,* 1(1): 1773–1778.
118. Esfandyari, M.J., et al. 2019. A hybrid model predictive and fuzzy logic based control method for state of power estimation of series-connected Lithium-ion batteries in HEVs. *Journal of Energy Storage* 24 (8): 100758–100769.
119. Park, J., et al. 2020. Integrated approach based on dual extended Kalman filter and multivariate autoregressive model for predicting battery capacity using health indicator and SOC/SOH. *Energies* 13 (9): 2138–2150.
120. Chu, Z., et al. 2019. A control-oriented electrochemical model for lithium-ion battery, Part I: Lumped-parameter reduced-order model with constant phase element. *Journal of Energy Storage* 25 (10): 100828–100839.
121. Gengfeng, L., Z. Xiangwen, and L. Zhiming. 2022. State of health estimation of power batteries based on multi-feature fusion models using stacking algorithm. *Energy* 259 (15): 124851–124865.
122. Chen, Z., et al. 2019. State of health estimation for lithium-ion batteries based on fusion of autoregressive moving average model and Elman neural network. *IEEE Access* 7 (1): 102662–102678.

Chapter 2
Electrochemical Modeling of Energy Storage Lithium-Ion Battery

Abstract This chapter first commences with a comprehensive elucidation of the fundamental charge and discharge reaction mechanisms inherent in energy storage lithium batteries. Then, based on the simplified conditions of the electrochemical model, a SP model considering the basic internal reactions, solid-phase diffusion, reactive polarization, and ohmic polarization of the SEI film in the energy storage lithium-ion battery is established. The open-circuit voltage of the model needs to be solved using a simplified solid-phase diffusion equation. In addition, based on the SP model, this chapter builds an ESP model considering the liquid-phase potential to improve the model accuracy. The liquid concentration polarization overpotential of ESP model also needs to be solved by simplifying the liquid diffusion equation. Finally, this chapter describes a multi-cell model of energy storage battery pack using the ESP model as a cell model, and presents the terminal voltage expression of the battery pack model.

Keywords Lithium-ion battery · Energy storage · SP model · ESP model · Terminal voltage expression · Battery pack · Multi-cell model

2.1 Working Principle Analysis of Energy Storage Batteries

In practical engineering applications, the type of lithium energy storage battery is lithium iron phosphate battery. The active material for the negative electrode of an energy storage lithium battery is generally graphite, petroleum coke, or amorphous carbon, while the active material for the positive electrode is lithium iron phosphate. The positive and negative current collectors of energy storage lithium batteries are generally made of aluminum and copper, respectively. The membrane of energy storage lithium batteries is generally made of polyolefin material, which has the function of isolating the transmission of positive and negative electrons but allowing the free passage of lithium-ions. The electrolyte for energy storage lithium batteries consists of solutes and solvents that can conduct ions. The battery shell is generally square or cylindrical, used to protect the internal materials of the battery.

Q. Huang et al., *Long-Term Health State Estimation of Energy Storage Lithium-Ion Battery Packs*, https://doi.org/10.1007/978-981-99-5344-8_2

The working mechanism of energy storage lithium batteries during charging and discharging is lithium-ion intercalation and de intercalation caused by redox reactions. During charging, the lithium iron phosphate on the positive electrode undergoes an oxidation reaction, and lithium-ions are removed from the electrolyte to generate electrons, which are transferred from the positive electrode to the negative electrode, resulting in the generation of current in the external circuit. The working mechanism of the discharge process is opposite to the charging process. During the charging and discharging process, the positive electrode, negative electrode, and total reaction chemistry equations of the energy storage lithium battery are shown in Eq. (2.1).

$$\begin{cases} \mathrm{P}: LiFePO_4 \underset{d}{\overset{c}{\longleftrightarrow}} Li_{(1-x)}FePO_4 + xLi^+ + xe^- \\ \mathrm{N}: 6C + xLi^+ + xe^- \underset{d}{\overset{c}{\longleftrightarrow}} Li_xC_6 \\ \mathrm{T}: LiFeO_4 + 6C \underset{d}{\overset{c}{\longleftrightarrow}} Li_{(1-x)}FePO_4 + Li_xC_6 \end{cases} \tag{2.1}$$

As can be seen from Eq. (2.1), when charging a lithium energy storage battery, the lithium-ions in the lithium iron phosphate crystal are removed from the positive electrode and transferred to the negative electrode. The new lithium-ion insertion process is completed through the free electrons generated during charging and the carbon elements in the negative electrode. The process during discharge is its reverse reaction. Specifically, the schematic diagram of the working mechanism of the energy storage lithium battery is shown in Fig. 2.1.

As shown in Fig. 2.1, during discharge, the negative electrode generates free electrons and flows through the load as its function. At this time, chemical energy is converted into electrical energy. In addition, lithium-ions removed from the negative electrode diffuse to the positive electrode due to concentration differences for re insertion. Therefore, during the discharge process, the positive electrode is rich in

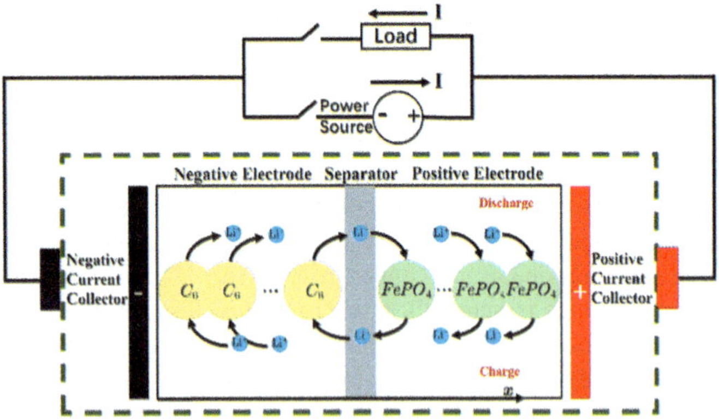

Fig. 2.1 Working principle of energy storage batteries

lithium and the negative electrode is poor in lithium. The energy conversion and reaction process during charging is opposite to that during discharge.

2.2 SP Modeling of Energy Storage Lithium Battery Considering the Influence of SEI Film

2.2.1 Research on the Simplification Mechanism of SP Model

Lithium-ion battery is a highly complex time-varying nonlinear electrochemical energy storage device, which is difficult to accurately describe the internal reaction mechanism [1]. Therefore, in order to balance computational complexity and model accuracy, how to select a suitable battery electrochemical model for construction is the key to effectively obtain health state estimation results [2]. The purpose of establishing energy storage lithium batteries in this book is to obtain electrochemical parameters closely related to battery health state. In Chap. 1, the P2D model of lithium-ion batteries and its typical simplified models have been reviewed in detail. It is mentioned that the SP simplified model has the characteristics of simple structure, low complexity, and poor simulation results in high magnification and complex operating conditions. However, energy storage lithium batteries often operate under low magnification conditions. Therefore, using the SP model can accurately characterize the internal electrochemical characteristics of energy storage lithium batteries within a certain range. The traditional SP model does not consider the terminal voltage changes caused by the SEI film, which leads to insufficient accuracy of the battery model. The structural diagram of the SP model for energy storage lithium-ion batteries considering the influence of SEI films is shown in Fig. 2.2.

Figure 2.2 is a schematic diagram of the SP model structure of an energy storage lithium iron phosphate battery. Where, x represents the electrode thickness direction, r represents the radial direction of active particles within the electrode, L_n, L_{sep}, and L_p represent the negative electrode thickness, separator thickness and positive

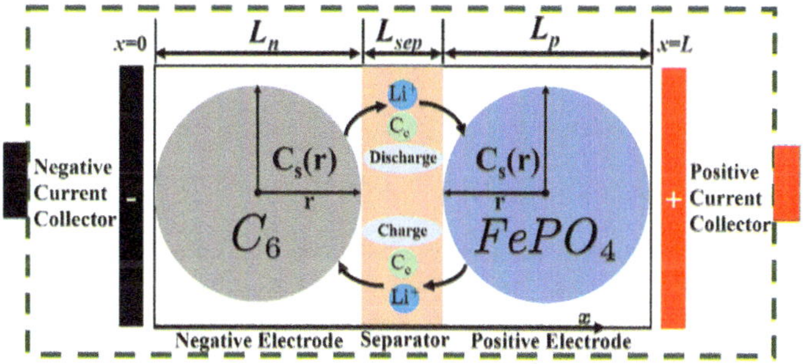

Fig. 2.2 The structural diagram of the SP model considering the influence of the SEI film

electrode thickness, respectively, and L represents the total thickness of the battery. By comparing Figs. 2.1 and 2.2, it can be seen that there is only one spherical particle at the positive and negative electrodes of the SP model, which is because the SP model uses the characteristics of a single spherical particle to characterize the characteristics of all particles in the electrode. The establishment of the SP model for energy storage lithium batteries requires the simplification of the following conditions based on the baseline electrochemical model [3–5]:

(1) It is believed that the spherical particles in the electrodes of energy storage lithium batteries are identical in shape and characteristics;
(2) It is believed that the current bulk density of the energy storage lithium-ion battery during operation is uniformly distributed among all active particles;
(3) It is believed that the concentration of lithium-ions in the liquid-phase in the internal structure of the energy storage lithium-ion battery is a constant, that is, the variation of the concentration of lithium-ions in the liquid-phase is not considered;
(4) It is believed that the internal electrochemical reaction of energy storage lithium-ion batteries is not affected by temperature changes during charging and discharging;
(5) It is believed that the potential difference between the solid particles in the electrode and the solid particles in the energy storage lithium battery is zero;
(6) It is believed that the terminal voltage of energy storage lithium-ion batteries is not affected by liquid voltage and liquid concentration, but only by open-circuit voltage, reactive polarization, and ohmic polarization.

From the above simplified conditions, it can be seen that the SP model assumes that the electrolyte concentration gradient and liquid phase diffusion do not exist when it is established. Therefore, the liquid phase diffusion phenomenon and the ohmic polarization process of solid particles and liquid electrolyte were ignored when the model was established. According to the internal physical characteristics of the energy storage lithium battery, the battery terminal voltage is the difference between the solid phase potential at the positive and negative electrode boundaries. Because the SP model ignores the liquid diffusion process, the liquid potential at each position in the electrode is zero. Based on the above requirements and the electrochemical characteristics of energy storage lithium batteries, the terminal voltage expression of the SP model is shown in Eq. (2.2).

$$\begin{aligned} U_t = \varphi_{s,p}(L) - \varphi_{s,n}(0) = E_p(\theta_p) - \eta_{act,p} - \eta_{SEI,p} \\ - \left[E_n(\theta_n) - \eta_{act,n} - \eta_{SEI,n}\right] = E - \eta_{act} - \eta_{SEI} \end{aligned} \tag{2.2}$$

In Eq. (2.2), U_t represents the terminal voltage, $\varphi_{s,p}$ represents the positive electrode solid-phase potential, $\varphi_{s,n}$ represents the negative electrode solid-phase potential, E represents the open-circuit voltage, E_p represents the positive electrode open circuit voltage, E_n represents the negative electrode open-circuit voltage, θ_p represents the utilization rate of the positive electrode, θ_n represents the utilization rate of the negative electrode, η_{act} represents the reactive polarization overpotential, $\eta_{act,p}$ represents the positive electrode reaction polarization overpotential, $\eta_{act,n}$ represents the negative electrode reaction polarization overpotential, η_{SEI} represents the ohmic polarization overpotential of the total SEI film, $\eta_{SEI,p}$ represents the ohmic polarization overpotential of the positive SEI film, $\eta_{SEI,n}$ represents the ohmic polarization overpotential of the negative SEI film. The overpotential is defined as the potential difference of the electrode.

2.2.2 *Solution of Open-Circuit Voltage Based on Solid-Phase Diffusion*

From Eq. (2.2), it can be seen that the open-circuit voltage is a function of θ_p and θ_n, and θ_i is an expression for the surface solid-phase lithium-ion concentration and the maximum solid-phase lithium-ion concentration in the electrode. Therefore, the specific expression of the open circuit voltage is shown in Eq. (2.3).

$$\begin{cases} E = E_p(\theta_p) - E_n(\theta_n) \\ \theta_p = \dfrac{c_{s,surf,p}}{c_{s,\max,p}}, \theta_n = \dfrac{c_{s,surf,n}}{c_{s,\max,n}} \end{cases} \tag{2.3}$$

Among them, $c_{s,surf,i}$ represents the surface solid-phase lithium-ion concentration, and $c_{s,max,i}$ represents the maximum solid-phase lithium-ion concentration, i represents the positive electrode or negative electrode.

The purpose of establishing an electrochemical model in this book is to obtain electrochemical parameters that have strong correlation with battery health state. According to the electrochemical mechanism, $c_{s,max,p}$ and $c_{s,max,n}$ are the most suitable parameter choices, and they cannot be directly obtained in most cases. Therefore, these two electrochemical parameters are usually obtained through identification methods after model construction. Therefore, in Eq. (2.3), the calculation expressions for $c_{s,max,p}$ and $c_{s,max,n}$ need to be established, which can be obtained by simplifying the solid phase diffusion process. Since the lithium-ion solid-phase diffusion process in the battery meets Fick's second law, the specific control procedure for solid phase diffusion is shown in Eq. (2.4).

$$\begin{cases} \dfrac{\partial c_{s,i}(t,r)}{\partial t} = D_{s,i,eff}\dfrac{1}{r^2}\dfrac{\partial}{\partial r}\left(r^2\dfrac{\partial c_{s,i}(t,r)}{\partial r}\right) \\ D_{s,i,eff} = D_{s,i}\varepsilon_{s,i}^{1.5} \end{cases} \tag{2.4}$$

In Eq. (2.4), $c_{s,i}(t,r)$ represents the concentration of lithium-ions in the solid-phase, $D_{s,i}$ represents the diffusion coefficient of lithium-ions in the solid-phase, $D_{s,i,eff}$ represents the effective diffusion coefficient of lithium-ions in the solid-phase, $\varepsilon_{s,i}$ represents the volume fraction of the solid-phase, r represents the radial distance coordinate of the solid-phase particles in the electrode, and t represents the time coordinate. Here, i can represent p, n, and sep, which respectively represent the positive electrode, negative electrode, and separator. In the following content of this book, the specific meaning of i will not be repeated. Due to the constant concentration at the center of the solid-phase particles, the two boundary conditions of Eq. (2.4) are shown in Eq. (2.5).

$$\begin{cases} \frac{\partial c_{s,i}(t,r)}{\partial r}|_{r=0} = 0 \\ \frac{\partial c_{s,i}(t,r)}{\partial r}\bigg|_{r=R_{s,i}} = -\frac{j_i}{D_{s,i}} \end{cases} \tag{2.5}$$

In Eq. (2.5), j_i represents the molar reaction flux density at the electrode current collector boundary, and $R_{s,i}$ represents the radius of solid-phase particles in the electrode. Based on the simplification conditions of the SP model, the expression of molar reaction flux density j_i at the positive and negative electrode current collector boundaries is shown in Eq. (2.6).

$$\begin{cases} j_i = \frac{I}{a_s F L_i A_i} \\ a_s = \frac{3\varepsilon_{s,i}}{R_{s,i}} \end{cases} \tag{2.6}$$

In Eq. (2.6), I represents the current passing through the battery, F represents the Faraday constant, L_i represents the thickness of an electrode, A_i represents the effective area of an electrode, a_s represents the surface area of spherical active particles per unit volume, and $\varepsilon_{s,i}$ represents the volume fraction of the solid-phase. The molar reaction flux density j_i at the positive and negative electrode current collector boundaries is positive for the negative electrode and negative for the positive electrode. Since the diffusion equation for the solid-phase concentration Eq. (2.4) is a parabolic partial differential equation, the three-parameter parabolic interpolation method is used to simplify the diffusion equation of the solid-phase concentration to solve for $c_{s,surf,i}$. After simplifying Eq. (2.4) accordingly, the distribution expression of the concentration of solid-phase lithium-ions is shown in Eq. (2.7).

$$c_{s,i}(t,r) = a(t) + b(t)\left(\frac{r^2}{R_{s,i}^2}\right) + c(t)\left(\frac{r^4}{R_{s,i}^4}\right) \tag{2.7}$$

In Eq. (2.7), $a(t)$, $b(t)$, and $c(t)$ are coefficients that need to be solved. Substituting Eq. (2.7) into the solid-phase diffusion equation shown in Eq. (2.4) and the boundary condition equation shown in Eq. (2.5) for solution yields a system of two equations about the coefficients $a(t)$, $b(t)$, and $c(t)$, as shown in Eq. (2.8).

$$\begin{cases} \dfrac{da(t)}{dt} = \dfrac{r^2}{R_{s,i}^2}\dfrac{db(t)}{dt} + \dfrac{r^4}{R_{s,i}^4}\dfrac{dc(t)}{dt} = \dfrac{2D_{s,i}}{R_{s,i}^2}\left(3b(t) + 10\dfrac{r^2}{R_{s,i}^2}c(t)\right) \\ \dfrac{2D_{s,i}}{R_{s,i}^2}b(t) + \dfrac{4D_{s,i}}{R_{s,i}}c(t) + j_i = 0 \end{cases} \tag{2.8}$$

For the solid-phase diffusion process in the active particles, the average concentration of solid-phase lithium-ions, the surface concentration of solid-phase lithium-ions, and the average concentration flux of particles are crucial. As Eq. (2.8) contains three unknowns, the expressions for the above three parameters are introduced for the solution of Eq. (2.8). The specific calculation process for the three parameters is shown in Eq. (2.9).

$$\begin{cases} q_{s,avg,i}(t) = \displaystyle\int_{r=0}^{R_{s,i}} 3\frac{r^2}{R_{s,i}^2}\left(\frac{\partial}{\partial r}c_{s,i}(t,r)\right)d\left(\frac{r}{R_{s,i}}\right) = \frac{2b(t)}{3R_{s,i}} + 2\frac{c(t)}{R_{s,i}} \\ c_{s,surf,i}(t) = a(t) + b(t) + c(t) \\ c_{s,avg,i}(t) = \displaystyle\int_{r=0}^{R_{s,i}} 3\frac{r^2}{R_{s,i}^2}c_{s,i}(t,r)d\left(\frac{r}{R_{s,i}}\right) = a(t) + \frac{3}{5}b(t) + \frac{3}{7}c(t) \end{cases} \tag{2.9}$$

In Eq. (2.9), $q_{s,avg,i}$ represents the average concentration flux of some electrode particles, $c_{s,surf,i}$ represents the surface lithium-ion concentration of the electrode in solid-phase, and $c_{s,avg,i}$ represents the average lithium-ion concentration of the electrode in solid-phase. By rearranging Eq. (2.9), expressions for the unknown variables $a(t)$, $b(t)$, and $c(t)$ can be obtained in terms of $q_{s,avg,i}$, $c_{s,surf,i}$, and $c_{s,avg,i}$, as shown in Eq. (2.10).

$$\begin{cases} a(t) = \dfrac{39}{4}c_{s,surf,i}(t) - 3q_{s,avg,i}(t)R_{s,i} - \dfrac{35}{4}c_{s,avg,i}(t) \\ b(t) = -35c_{s,surf,i}(t) + 10q_{s,avg,i}(t)R_{s,i} + 35c_{s,avg,i}(t) \\ c(t) = \dfrac{105}{4}c_{s,surf,i}(t) - 7q_{s,avg,i}(t)R_{s,i} - \dfrac{105}{4}c_{s,avg,i}(t) \end{cases} \tag{2.10}$$

By substituting the expressions for $a(t)$, $b(t)$, and $c(t)$ Eq. (2.10) into Eq. (2.7), an expression for $c_{s,max,i}$ within the positive and negative electrodes is obtained, as shown in Eq. (2.11).

$$\begin{aligned} c_{s,i}(t,r) =& \frac{39}{4}c_{s,surf,i}(t) - 3q_{s,avg,i}(t)R_{s,i} - \frac{35}{4}c_{s,avg,i}(t) \\ &+ \left[-35c_{s,surf,i}(t) + 10q_{s,avg,i}(t)R_{s,i} + 35c_{s,avg,i}(t)\right]\left(\frac{r^2}{R_{s,i}^2}\right) \end{aligned}$$

$$+\left[\frac{105}{4}c_{s,surf,i}(t)-7q_{s,avg,i}(t)R_{s,i}-\frac{105}{4}c_{s,avg,i}(t)\right]\left(\frac{r^4}{R_{s,i}^4}\right) \tag{2.11}$$

By combining the expression for solid-phase lithium-ion concentration shown in Eq. (2.11), the boundary condition expressions shown in Eq. (2.5), and the integral equation for average solid-phase lithium-ion concentration and particle average concentration flux shown in Eq. (2.9), expressions for calculating the average solid-phase lithium-ion concentration and particle average concentration flux can be obtained, as shown in Eq. (2.12).

$$\begin{cases} c_{s,avg,i}(t)=c_{s,0,i}-\int\limits_0^t 3\frac{j_i}{R_{s,i}}dt \\ \frac{d}{dt}q_{s,avg,i}(t)+30\frac{D_{s,i}}{R_{s,i}^2}q_{s,avg,i}(t)+\frac{45j_i}{2R_{s,i}^2}=0 \end{cases} \tag{2.12}$$

In Eq. (2.12), $c_{s,0,i}$ represents the initial solid-phase lithium-ion concentration. For ease of calculation and analysis during modeling, the expression for particle average concentration flux in Eq. (2.12) is often discretized. The resulting iterative expression for particle average concentration flux is shown in Eq. (2.13).

$$\begin{cases} constant\ discharge: q_{s,avg,i}(t)=\frac{3j_i}{4D_{s,i}}\left[\exp(-t/\tau_i^s)-1\right],\ \tau_i^s=\frac{R_{s,i}^2}{30D_{s,i}} \\ other\ conditions: q_{s,avg,i}(t_{k+1})=q_{s,avg,i}(t_k)-\left[\frac{45j_i}{2R_{s,i}^2}+\frac{q_{s,avg,i}(t_k)}{\tau_i^s}\right](t_{k+1}-t_k) \end{cases} \tag{2.13}$$

The algebraic equation in Eq. (2.8) consists of constants and expressions for $a(t)$ and $b(t)$, both of which are expressions involving the three important parameters mentioned above. The expression for surface solid-phase lithium-ion concentration is thus given by Eq. (2.14).

$$c_{s,surf,i}(t)=c_{s,avg,i}(t)+\left(8D_{s,i}q_{s,avg,i}(t)-j_i\right)\frac{R_{s,i}}{35D_{s,i}} \tag{2.14}$$

Equations (2.12) and (2.14) represent the simplified solid-phase diffusion process of the storage lithium-ion battery SP model through the three-parameter parabolic method. In Eqs. (2.12), (2.13), and (2.14), the boundary conditions for particle average concentration flux and surface solid-phase lithium-ion concentration can be described as shown in Eq. (2.15).

$$\begin{cases} q_{s,avg,i}(t)|_{t=0}=0 \\ c_{s,surf,i}|_{t=0}=c_{s,surf,0,i} \end{cases} \tag{2.15}$$

In Eq. (2.15), $c_{s,surf,0,i}$ represents the surface solid-phase lithium-ion concentration at the initial time of the new battery. By combining Eqs. (2.12), (2.14), and the relevant boundary conditions Eq. (2.15), an expression for the positive and negative electrode surface solid-phase lithium-ion concentration is solved, thereby making the unknown parameter in the open-circuit voltage expression shown in Eq. (2.3) only the maximum solid-phase lithium-ion concentrations of the positive and negative electrodes.

2.2.3 Solution of Reactive Polarization Overpotential Based on Butler-Volmer Dynamics

In Eq. (2.2), η_{act} represents the reaction polarization overpotential, which is an important factor affecting the reaction rate and efficiency. The expression for calculating the reactive polarization overpotential is shown in Eq. (2.16).

$$\eta_{act} = \eta_{act,n} - \eta_{act,p} \tag{2.16}$$

Traditional methods for solving reaction polarization overpotential are mostly based on electrochemical kinetic theory, but this method requires a high understanding of the reaction mechanism and charge transfer process, and is also subject to many limitations. Therefore, this book proposes a reaction polarization overpotential solution method based on Butler-Volmer dynamics, which can simplify the solution process and has wider applicability. A kinetic model of the reaction process was established using Butler-Volmer kinetic theory, including factors such as reactant diffusion, charge transfer, and reaction rate. The Butler-Volmer expression is shown in Eq. (2.17).

$$\begin{cases} j_i = j_0\left(\exp\left(\dfrac{0.5F}{RT}\right)\eta_{act,i} - \exp\left(-\dfrac{0.5F}{RT}\right)\eta_{act,i}\right) \\ j_0 = k_i\left(c_{s,\max,i} - c_{s,surf,i}(t)\right)^{0.5}\left(c_{s,surf,i}(t)\right)^{0.5} c_e^{0.5} \end{cases} \tag{2.17}$$

In Eq. (2.17), k_i represents the average electrochemical reaction rate constant of the electrode, c_e represents the concentration of lithium ions in the liquid-phase, R represents the universal gas constant, and T represents the average internal temperature of the battery. By solving Eq. (2.17), the expressions of the positive reaction polarization overpotential and the negative reaction polarization overpotential are shown in Eq. (2.18).

$$\begin{cases} \eta_{act,i} = \dfrac{2RT}{F}\ln\left(m_i + \sqrt{m_i^2 + 1}\right) \\ m_i = \dfrac{j_i}{2k_i\left(c_{s,\max,i} - c_{s,surf,i}(t)\right)^{0.5}\left(c_{s,surf,i}(t)\right)^{0.5} c_e^{0.5}} \end{cases} \tag{2.18}$$

After obtaining the positive reaction polarization overpotential and the negative reaction polarization overpotential shown in Eq. (2.18), combined with the internal relationship expression of reaction polarization overpotential shown in Eq. (2.16), the reaction polarization overpotential can be calculated.

2.2.4 *Solution of SEI Film Ohmic Polarization Overpotential Based on Ohmic Law*

The SP model of lithium-ion batteries constructed in this book considers the influence of SEI films on the modeling process. Therefore, in this model, the ohmic polarization overpotential is the SEI film ohmic polarization overpotential, and its calculation process is shown in Eq. (2.19).

$$\eta_{SEI} = \eta_{SEI,n} - \eta_{SEI,p} = R_{SEI,n}Fj_n - R_{SEI,p}Fj_p \tag{2.19}$$

In Eq. (2.19), $R_{SEI,p}$ represents the SEI film ohmic resistance on the positive electrode, and $R_{SEI,n}$ represents the SEI film ohmic resistance on the negative electrode. Combining the expression for terminal voltage in the SP model, the expression for open-circuit voltage, the simplified expressions for solid-phase diffusion, the expression for reaction polarization overpotential, and the expression for ohmic polarization overpotential, the specific calculation formula for the terminal voltage in the SP model of the energy storage lithium-ion battery is obtained, as shown in Eq. (2.20).

$$U_t = E_p(\theta_p) - E_n(\theta_n) - \frac{2RT}{F}\frac{\ln\left(m_n + \sqrt{m_n^2 + 1}\right)}{\ln\left(m_p + \sqrt{m_p^2 + 1}\right)} - \left(R_{SEI,n}Fj_n - R_{SEI,p}Fj_p\right) \tag{2.20}$$

Based on Eq. (2.20), during the charging-discharging process of the energy storage lithium-ion battery, the current can be taken as the input and the terminal voltage as the output to establish the SP model, and the required internal electrochemical parameters of the battery can be solved according to this model. The block diagram of the SP model is shown in Fig. 2.3.

As can be seen from Fig. 2.3, the block diagram of the SP model includes two reaction modules: positive electrode and negative electrode. The positive and negative electrode modules contain two graphs that represent the change of the positive electrode open circuit voltage with the electrode positive electrode utilization rate and

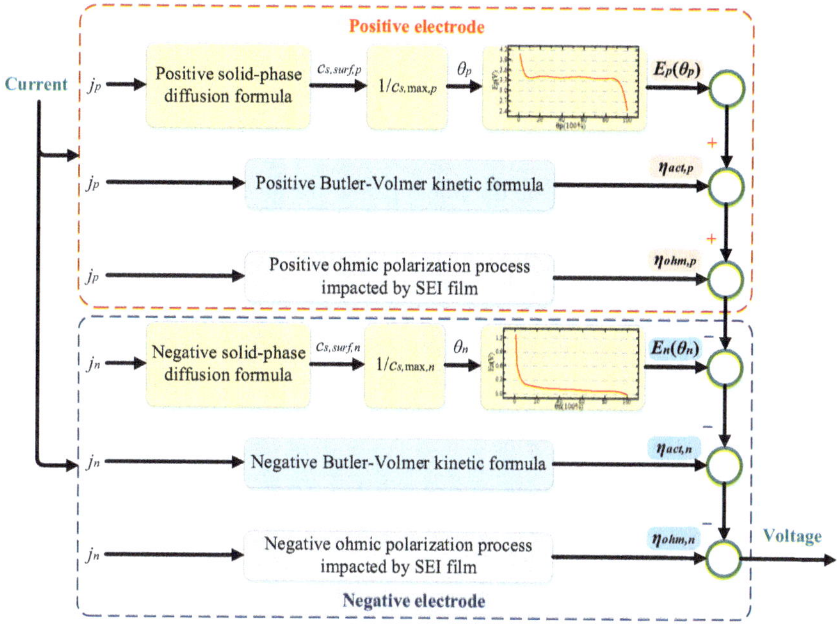

Fig. 2.3 The block diagram of the SP model

the change of the negative electrode open circuit voltage with the electrode negative electrode utilization rate, respectively.

2.3 ESP Modeling of Energy Storage Lithium Battery Considering Liquid-Phase Potential

2.3.1 Research on Optimization and Improvement of ESP Model

The operating current of an energy storage lithium battery will generate a sudden variable at the moment of stopping discharging. However, due to the imperfect consideration of the battery ohmic internal resistance in the SP model built in the above research, there is a large error between the terminal voltage sudden variable corresponding to the simulation in the SP model and the actual terminal voltage sudden variable [6–8]. In order to reduce the model error caused by this problem, the ESP model established in this book considers the liquid potential caused by liquid concentration polarization and liquid ohmic polarization in energy storage lithium batteries based on the SP model. Therefore, the expressions for the liquid phase potential inside the battery and the ESP model terminal voltage are shown in Eq. (2.21).

$$\begin{cases} \eta_{liq} = \eta_{liq,n} - \eta_{liq,p} = \eta_{con-pol,n} + \eta_{liq-ohm,n} - \left(\eta_{con-pol,p} + \eta_{liq-ohm,p}\right) \\ = \eta_{con-pol,n} - \eta_{con-pol,p} + \eta_{liq-ohm,n} - \eta_{liq-ohm,p} = \eta_{con-pol} + \eta_{liq-ohm} \\ U_t = E - \eta_{act} - \eta_{SEI} - \eta_{liq} \end{cases} \tag{2.21}$$

In Eq. (2.21), η_{liq} represents the liquid-phase overpotential, $\eta_{liq,i}$ represents the liquid-phase overpotential in the electrode i, $\eta_{con\text{-}pol}$ represents the liquid-phase concentration polarization overpotential, $\eta_{liq\text{-}ohm}$ represents the liquid-phase ohmic polarization overpotential, $\eta_{con\text{-}pol,i}$ represents the liquid-phase concentration polarization overpotential in the electrode i, and $\eta_{liq\text{-}ohm,i}$ represents the liquid-phase ohmic polarization overpotential in the electrode i. The liquid-phase potential includes the liquid-phase concentration polarization overpotential and the liquid-phase ohmic polarization overpotential. Therefore, the liquid-phase potential distribution of electrodes and the separator in the battery is shown in Eq. (2.22).

$$\begin{cases} \dfrac{\partial \varphi_{e,i}}{\partial x} = -\dfrac{i_{e,i}}{\kappa_{eff,i}} + \dfrac{2RT}{F}(1 - t_+)\dfrac{\partial \ln c_{e,i}}{\partial x} \\ \kappa_{eff,i} = \kappa_i \varepsilon_{e,i}^{1.5} \end{cases} \tag{2.22}$$

The first term on the right side of Eq. (2.22) is based on the liquid-phase ohmic law, describing the change in liquid-phase potential caused by the liquid-phase ohmic internal resistance. In addition, the second term on the right of Eq. (2.22) describes the liquid-phase potential change caused by the liquid-phase lithium-ion concentration difference. Besides, $\varphi_{e,i}$ represents the liquid-phase potential, $i_{e,i}$ represents the liquid-phase current density, κ_i represents the liquid-phase lithium-ion conductivity, $\kappa_{eff,i}$ represents the effective liquid-phase lithium-ion conductivity, $\varepsilon_{e,i}$ represents the liquid-phase volume fraction, x represents the electrode thickness coordinate from the negative electrode to the positive electrode of the battery, R represents the molar gas constant, T represents the internal instantaneous temperature, F represents the Faraday constant, c_e represents the liquid-phase lithium-ion concentration, and t_+ represents the liquid-phase lithium-ion transfer coefficient. The expressions for boundary conditions Eq. (2.22) are shown in Eq. (2.23).

$$\begin{cases} \dfrac{\partial \varphi_e}{\partial x}|_{x=0} = \dfrac{\partial \varphi_e}{\partial x}|_{x=L} = 0 \\ \varphi_e\big|_{x=L_n^-} = \varphi_e\big|_{x=L_n^+}, \varphi_e\big|_{x=L_n+L_{sep}^-} = \varphi_e\big|_{x=L_n+L_{sep}^+} \end{cases} \tag{2.23}$$

The first expression of Eq. (2.23) indicates that the liquid-phase potential at the current collector interface between the positive and negative electrodes is constant during the normal operation of the energy storage lithium battery, and the latter two expressions indicate that the electrode liquid-phase potential and the separator liquid-phase potential are continuous at the interface during the normal operation of the energy storage lithium battery.

2.3.2 Solution of Liquid-Phase Concentration Polarization Overpotential Based on Liquid-Phase Concentration Distribution

When solving the concentration polarization overpotential in the liquid-phase potential based on Eq. (2.22), the liquid-phase ohmic polarization overpotential is not considered. Therefore, the concentration polarization overpotential in the liquid-phase potential can be expressed from Eq. (2.22) and the distribution of lithium ions in the electrodes, as shown in Eq. (2.24).

$$\eta_{con-pol} = \eta_{con-pol,n} - \eta_{con-pol,p} = \frac{2RT}{F}(1 - t_+) \ln \frac{c_e(0)}{c_e(L)} \tag{2.24}$$

It can be seen from Eq. (2.24) that to obtain the concentration polarization overpotential in the liquid-phase potential, it is necessary to calculate the liquid-phase lithium-ion concentration at the electrode boundaries in advance. The main reason for the uneven concentration of lithium ions in the liquid-phase is the phenomenon that lithium ions are deintercalated into the electrolyte during the solid phase diffusion of active particles, resulting in a local increase in the concentration of lithium ions in the liquid-phase here. The uneven concentration of lithium ions in the liquid-phase leads to diffusion and migration of lithium ions in the liquid-phase caused by concentration differences, and the impact of the migration process on the battery reaction can be negligible. Thence, only the liquid-phase diffusion process is considered, and the specific expression is shown in Eq. (2.25).

$$\begin{cases} \varepsilon_{e,i} \dfrac{\partial c_{e,i}(t,x)}{\partial t} = D_{e,eff} \dfrac{\partial^2 c_{e,i}(t,x)}{\partial x^2} + a_s(1 - t_+) j_i \\ D_{e,eff} = D_e \varepsilon_{e,i}^{1.5} \end{cases} \tag{2.25}$$

In Eq. (2.25), $c_{e,i(t,x)}$ represent the concentration of lithium ions in the liquid-phase, D_e represents the diffusion coefficient of lithium ions in the liquid-phase, $D_{e,eff}$ represents the effective diffusion coefficient of lithium ions in the liquid-phase, $\varepsilon_{e,i}$ represents the volume fraction of the liquid-phase, x represents the electrode thickness coordinate from the negative electrode to the positive electrode of the battery, and t represents the time coordinate. In addition, the first term on the right of Eq. (2.25) describes the change in lithium-ion concentration caused by liquid-phase concentration diffusion, and the second term on the right describes the change in lithium-ion concentration caused by electrochemical reactions at the interface between solid-phase particles and liquid-phase electrolyte. The initial and boundary conditions of Eq. (2.25) are shown in Eq. (2.26).

$$\begin{cases} c_e(t_0) = c_{e,0} \\ \dfrac{\partial c_e(t,x)}{\partial x}\Big|_{x=0} = \dfrac{\partial c_e(t,x)}{\partial x}\Big|_{x=L} = 0 \\ \dfrac{\partial c_e(t,x)}{\partial x}\Big|_{x=L_n^-} = \dfrac{\partial c_e(t,x)}{\partial x}\Big|_{x=L_n^+}, c_e\Big|_{x=L_n^-} = c_e\Big|_{x=L_n^+} \\ \dfrac{\partial c_e(t,x)}{\partial x}\Big|_{x=(L_n+L_{sep})^-} = \dfrac{\partial c_e(t,x)}{\partial x}\Big|_{x=(L_n+L_{sep})^+}, c_e\Big|_{x=(L_n+L_{sep})^-} = c_e\Big|_{x=(L_n+L_{sep})^+} \end{cases} \tag{2.26}$$

In Eq. (2.26), the expression in the first row describes the initial state of normal operation of the energy storage lithium battery, with the liquid-phase concentration in the internal region of the battery unchanged. The expression in the second row describes that there is no diffusion of lithium ions in the liquid-phase at the positive and negative current collectors of the energy storage lithium-ion battery, that is, the concentration gradient in the liquid-phase disappears. The expressions in the third and fourth rows describe the continuity of the liquid-phase concentration at the junction of the positive and negative electrodes of the battery and the separator, respectively.

When the energy storage lithium-ion battery reaches a stable state, the entry and exit of lithium ions from the solid-phase particles into the electrolyte is balanced due to the electrochemical competition effect and concentration gradient diffusion effect, that is, the left term of the first row of Eq. (2.25) is set to zero. In addition, in the steady state, the current distribution in the battery is uniform, and the second term on the right of the first row of Eq. (2.25) is set to be constant. Therefore, by substituting the molar reaction flux density expression at the electrode current collector boundary shown in Eq. (2.26) into the liquid-phase diffusion equation at the steady state, the liquid-phase lithium-ion concentration expression for the electrode and separator at the steady state can be obtained. In this book, the concentration of lithium ions in the liquid-phase at the separator is not required. Therefore, the distribution expressions of lithium-ion concentration in the positive and negative electrode liquid-phase are shown in Eq. (2.27).

$$c_{e,steady}(x) = \begin{cases} \left(\dfrac{P_1 x^2}{2} + P_2\right)\dfrac{I}{A_n} + c_{e,0}, \text{ (negative)} \\ \left[\dfrac{P_3 x(x-L)^2}{2} + P_4\right]\dfrac{I}{A_p} + c_{e,0}, \text{ (positive)} \end{cases} \tag{2.27}$$

In Eq. (2.27), P_1, P_2, P_3 and P_4 are constant coefficients that can be calculated from multiple parameters within the energy storage lithium battery. The specific expressions of them are shown in Eq. (2.28).

$$\begin{cases} P_1 = \dfrac{(1-t_+)}{D_e F}\dfrac{-1}{\varepsilon_{e,n}^{1.5} L_n} \\ P_2 = \dfrac{(1-t_+)}{D_e F}\dfrac{\frac{L_n^2}{6\varepsilon_{e,n}^{0.5}} + \frac{L_{sep}^2}{2\varepsilon_{e,sep}^{0.5}} + \frac{L_p^2}{3\varepsilon_{e,p}^{0.5}} + \frac{\varepsilon_{e,sep} L_n L_{sep}}{2\varepsilon_{e,n}^{1.5}} + \frac{\varepsilon_{e,p} L_{sep} L_p}{\varepsilon_{e,sep}^{1.5}} + \frac{\varepsilon_{e,p} L_n L_p}{2\varepsilon_{e,n}^{1.5}}}{\varepsilon_{e,n} L_n + \varepsilon_{e,sep} L_{sep} + \varepsilon_{e,p} L_p} \\ P_3 = \dfrac{(1-t_+)}{D_e F}\dfrac{-1}{\varepsilon_{e,p}^{1.5} L_p} \\ P_4 = \dfrac{(1-t_+)}{D_e F}\left(\dfrac{L_n}{2\varepsilon_{e,n}^{1.5}} + \dfrac{L_{sep}}{\varepsilon_{e,sep}^{1.5}} + \dfrac{L_p}{2\varepsilon_{e,p}^{1.5}}\right) \end{cases} \tag{2.28}$$

From Eq. (2.27) for the liquid-phase lithium-ion concentration at the steady state, it can be seen that the parabolic distribution of the liquid-phase lithium-ion concentration within the electrodes is related to the relevant time-varying coefficients. Therefore, in combination with Eq. (2.27), the concentration distribution of lithium ions in the liquid-phase inside the electrodes is shown in Eq. (2.29).

$$c_e(x,t) = \begin{cases} \left(\dfrac{P_1 x^2}{2} + P_2\right) f_n(t)\dfrac{I}{A_n} + c_{e,0}, \text{ (negative)} \\ \left[\dfrac{P_3 x (x-L)^2}{2} + P_4\right] f_p(t)\dfrac{I}{A_p} + c_{e,0}, \text{ (positive)} \end{cases} \tag{2.29}$$

In Eq. (2.29), $f_i(t)$ represents the time-varying coefficients corresponding to the parabolic distribution of lithium-ion concentration in the liquid-phase of the electrode i. During constant current discharge process, the continuous and discrete expressions of $f_i(t)$ are shown in Eq. (2.30).

$$\begin{cases} f_n(t) = 1 - \exp(-t/\tau_n),\ \tau_n = \dfrac{-1}{D_e \varepsilon_{e,n}^{0.5}}\dfrac{P_2}{P_1},\ f_p(t) = 1 - \exp(-t/\tau_p),\ \tau_p = \dfrac{-1}{D_e \varepsilon_{e,p}^{0.5}}\dfrac{P_4}{P_3} \\ f_n(t_{k+1}) = f_n(t_k) + \dfrac{1 - f_n(t_k)}{\tau_n}(t_{k+1} - t_k),\ f_p(t_{k+1}) = f_p(t_k) + \dfrac{1 - f_p(t_k)}{\tau_p}(t_{k+1} - t_k) \end{cases} \tag{2.30}$$

Combining Eqs. (2.29) and (2.30) with x equals zero and x equals L, the concentration of lithium ions in the liquid-phase at the positive and negative electrode boundaries can be obtained, as shown in Eq. (2.31).

$$\begin{cases} c_{e,n} = c_e(0) = P_2\left[1 - \exp\left(\dfrac{-t}{\tau_n}\right)\right]\dfrac{I}{A_n} + c_{e,0},\ \tau_n = \dfrac{-1}{D_e \varepsilon_{e,n}^{0.5}}\dfrac{P_2}{P_1} \\ c_{e,p} = c_e(L) = P_4\left[1 - \exp\left(\dfrac{-t}{\tau_p}\right)\right]\dfrac{I}{A_p} + c_{e,0},\ \tau_p = \dfrac{-1}{D_e \varepsilon_{e,p}^{0.5}}\dfrac{P_4}{P_3} \end{cases} \tag{2.31}$$

After obtaining the concentration of lithium ions in the liquid-phase at the positive and negative electrode boundaries, the concentration polarization overpotential in the liquid phase potential can be solved by according to Eq. (2.24).

2.3.3 Solution of Liquid-Phase Ohmic Polarization Overpotential Based on Liquid-Phase Ohmic Law

When solving the liquid-phase ohmic polarization overpotential based on the liquid-phase potential distribution expression shown in Eq. (2.22), the concentration polarization overpotential is not considered. Therefore, the liquid-phase ohmic polarization overpotential can be calculated from the effective conductivity of the liquid-phase lithium-ion and the liquid-phase current density, as shown in Eq. (2.32).

$$\begin{cases} \frac{\partial \varphi_{e,i}}{\partial x} = -\frac{i_{e,i}}{\kappa_{eff,i}} \\ \kappa_{eff,i} = \kappa_i \varepsilon_{e,i}^{1.5} \end{cases} \tag{2.32}$$

Equation (2.32) meets liquid-phase ohmic law. In Eq. (2.32), $\kappa_{eff,i}$ is the effective conductivity of the liquid-phase lithium-ion, and i_e is the liquid-phase current. In addition, because the change of liquid-phase current density depends on the rate of lithium-ion deintercalation, the expression of the change rate of liquid-phase current density can be obtained based on the charge conservation principle, which is shown in Eq. (2.33).

$$\frac{\partial i_e(x)}{\partial x} = a_s F j_i = \frac{I}{L_i A_i} \tag{2.33}$$

According to Eq. (2.33) and the average local volume current density at the electrodes, the liquid-phase current density of the positive and negative electrodes can be expressed as Eq. (2.34).

$$\begin{cases} i_{e,n}(x) = i_e(0) + \int_0^x \frac{\partial i_e(x)}{\partial x} dx = \frac{Ix}{L_n A_n} \\ i_{e,p}(x) = i_e\left(L_n + L_{sep}\right) + \int_{L_n+L_{sep}}^x \frac{\partial i_e(x)}{\partial x} dx = \frac{I}{A_p} - \frac{I\left(x - L_n - L_{sep}\right)}{L_p A_p} \end{cases} \tag{2.34}$$

In Eq. (2.34), $i_{e,n}$ represents the negative electrode liquid-phase current density, and $i_{e,p}$ represents the positive electrode liquid-phase current density. Since the concentration polarization overpotential is not considered here, the positive and negative liquid-phase ohmic polarization overpotential can be expressed by combining Eqs. (2.32) and (2.34), as shown in Eq. (2.35).

$$\begin{cases} \eta_{liq-ohm,n}(x) = -\dfrac{Ix^2}{L_n A_n \kappa_{eff,n}} \\ \eta_{liq-ohm,p}(x) = -\dfrac{I}{2A_n}\left(\dfrac{L_n}{\kappa_{eff,n}} + \dfrac{2L_{sep}}{\kappa_{eff,sep}} + \dfrac{\left(x - L_n - L_{sep}\right)^2}{L_p \kappa_{eff,p}}\right) \end{cases} \tag{2.35}$$

The liquid-phase ohmic polarization overpotential inside the battery is the difference between the positive liquid-phase ohmic polarization overpotential and the negative liquid-phase ohmic polarization overpotential, and x is set to L and zero at the positive and negative electrode boundaries, respectively. Thence, the expression of liquid-phase ohmic polarization overpotential is shown in Eq. (2.36).

$$\eta_{liq-ohm} = \eta_{liq-ohm,n}(0) - \eta_{liq-ohm,p}(L) = \frac{I}{2A_n}\left(\frac{L_n}{\kappa_{eff,n}} + \frac{2L_{sep}}{\kappa_{eff,sep}} + \frac{L_p}{\kappa_{eff,p}}\right) \tag{2.36}$$

From the SP model voltage expression considering the influence of SEI film, liquid-phase concentration polarization overpotential expression, and liquid-phase ohmic polarization overpotential expression, it can be seen that the terminal voltage expression of the ESP model considering the liquid-phase potential is shown in Eq. (2.37).

$$\begin{aligned} U_t =& E_p(\theta_p) - E_n(\theta_n) - \frac{2RT}{F}\frac{\ln\left(m_n + \sqrt{m_n^2+1}\right)}{\ln\left(m_p + \sqrt{m_p^2+1}\right)} - \left(R_{SEI,n}Fj_n - R_{SEI,p}Fj_p\right) \\ & - \left[\frac{2RT}{F}(1-t_+)\ln\frac{c_e(0)}{c_e(L)} + \frac{I}{2A_n}\left(\frac{L_n}{\kappa_{eff,n}} + \frac{2L_{sep}}{\kappa_{eff,sep}} + \frac{L_p}{\kappa_{eff,p}}\right)\right] \end{aligned} \tag{2.37}$$

Equation (2.37) is the simulation terminal voltage expression of the ESP cell model for energy storage lithium-ion batteries with current as input, terminal voltage as output, and electrochemical parameters as internal state variables. Similar to the block diagram of the SP model described in Fig. 2.3, after considering the liquid-phase concentration distribution and the liquid-phase Ohmic law on the basis of the SP model, the block diagram of the ESP model considering the liquid-phase potential is constructed when the energy storage lithium-ion battery is in the normal operation, as shown in Fig. 2.4.

Comparing Figs. 2.3 and 2.4, it can be seen that the block diagram of the ESP model increases the impact of the liquid-phase potential of the positive and negative electrodes on the battery model compared to the block diagram of the SP model. The other content in the positive and negative electrode module of the ESP model block diagram is consistent with the structural block diagram of the SP model.

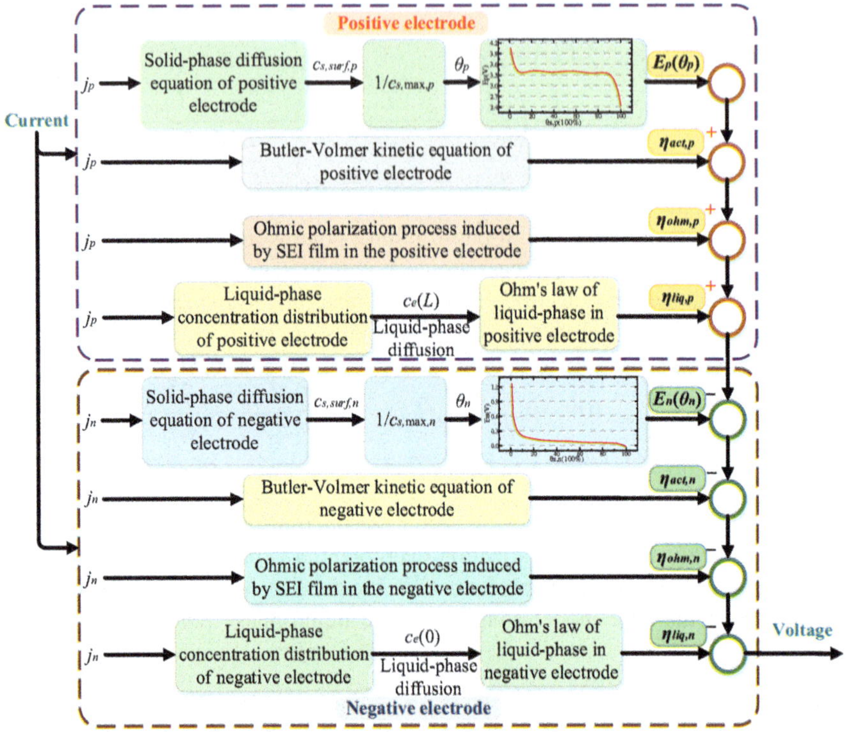

Fig. 2.4 The block diagram of the ESP model

2.4 Extended Single Particle-Multi Cell Modeling of Energy Storage Lithium-Ion Battery Pack

Effective health indicators are the key to accurate health state estimation model results. Health indicators include the maximum solid-phase lithium-ion concentrations of the positive and negative electrodes in the ESP model. However, due to inconsistencies between the cells in the battery pack, the values of health indicators in each cell are different. To solve this problem, it is crucial to establish a battery pack model that can independently describe the electrochemical reaction process of each cell. This book proposes a battery pack multi-cell model (MCM) based on the ESP model as a cell model. This pack model can meet the individual identification of electrochemical parameters, including the maximum solid-phase lithium-ion concentrations of each cell's positive and negative electrodes. It is worth noting that although MCM can be modeled separately using the parameters of each cell, it is only applicable to the battery pack with a series structure. The structure diagram of the extended single particle-multi cell model (ESP-MCM) based on the ESP model is shown in Fig. 2.5.

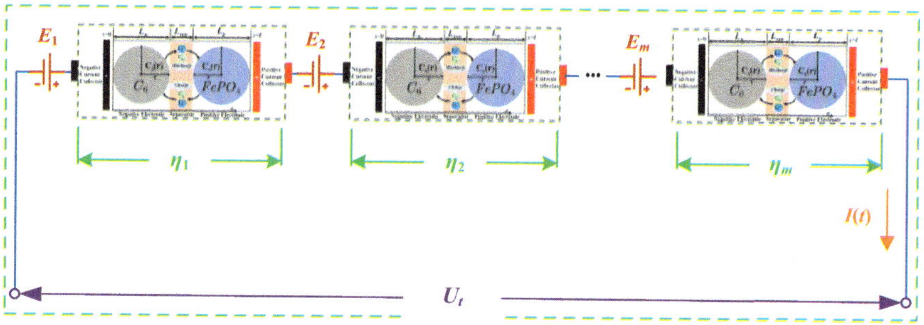

Fig. 2.5 The structure diagram of the ESP-MCM

In the proposed ESP-MCM, the terminal voltage of the cell m can be obtained from the ESP model terminal voltage output expression in Eq. (2.21), as shown in Eq. (2.38).

$$\begin{cases} U_{t,m} = E_m - \eta_m \\ \eta_m = \eta_{act,m} + \eta_{SEI,m} + \eta_{liq,m} \end{cases} \tag{2.38}$$

In Eq. (2.38), for the cell m in the ESP-MCM, E_m, η_m, $U_{t,m}$, $\eta_{act,m}$, $\eta_{SEI,m}$ and $\eta_{liq,m}$ represent the open-circuit voltage, the sum of all the overpotentials, the terminal voltage, the reaction polarization overpotential, the SEI film ohmic polarization overpotential and the liquid-phase ohmic polarization overpotential, respectively. Their specific calculation expressions are described in detail above. The terminal voltage expression of the series battery pack ESP-MCM is shown in Eq. (2.39).

$$U_t = \sum_{m=1}^{n} U_{t,m} \tag{2.39}$$

In Eq. (2.39), n represents the number of cells in the series battery pack ESP-MCM. According to the definition of series system, the terminal voltage of the battery pack ESP-MCM is the sum of the terminal voltages of each cell.

2.5 Summary

Considering the intricacy of energy storage lithium-ion batteries during their operation in real energy storage conditions, it becomes crucial to devise a battery model that exhibits engineering-suitable characteristics while elucidating internal degradation phenomena. To fulfill these requisites, an ESP cell model is established, which accounts for the liquid-phase potential. Building upon this framework, a MCM for the battery pack is developed, enabling comprehensive estimation of the health state for individual cells as well as the entirety of the pack.

References

1. Keil, et al. 2016. Calendar aging of lithium-ion batteries I. Impact of the graphite anode on capacity fade. *Journal of the Electrochemical Society*, 163 (9): 1872–1884.
2. Mastali, M., et al. 2018. Electrochemical-thermal modeling and experimental validation of commercial graphite/$LiFePO_4$ pouch lithium-ion batteries. *International Journal of Thermal Sciences* 129 (7): 218–230.
3. Li, J., et al. 2018. A single particle model with chemical/mechanical degradation physics for lithium ion battery State of Health (SOH) estimation. *Applied Energy* 212 (15): 1178–1190.
4. Grandjean, T. R. B., L. Li, M. X. Odio, and W. D. Widanage. 2019. Global sensitivity analysis of the single particle lithium-ion battery model with electrolyte. *IEEE Vehicle Power and Propulsion Conference (VPPC)*, 1 (1): 1–7.
5. Li, J., R.G. Landers, and J. Park. 2020. A comprehensive single-particle-degradation model for battery state-of-health prediction. *Journal of Power Sources* 456 (30): 227950–227962.
6. Rohit, M., and G. Amit. 2021. An improved single-particle model with electrolyte dynamics for high current applications of lithium-ion cells. *Electrochimica Acta* 389 (1): 138623–138631.
7. Feng, F., et al. 2022. Electrochemical impedance characteristics at various conditions for commercial solid–liquid electrolyte lithium-ion batteries: Part 1. experiment investigation and regression analysis. *Energy*, 242 (1): 122880–122890.
8. Feng, F., et al. 2022. Electrochemical impedance characteristics at various conditions for commercial solid–liquid electrolyte lithium-ion batteries: Part. 2. Modeling and prediction. *Energy*, 243 (1): 123091–123104.

Chapter 3
Extraction of Multidimensional Health Indicators Based on Lithium-Ion Batteries

Abstract The purpose of this chapter is to extract health indicators strongly related to the health state of energy storage lithium-ion batteries, including the maximum solid-phase lithium-ion concentrations in the positive and negative electrodes, the loss of active material (LAM), and the loss of lithium inventory (LLI). This chapter firstly describes the basic structure of an improved cooperation competition particle swarm optimization (CCPSO) algorithm, and then studies the identification process of important electrochemical parameters in the ESP model using this algorithm. The important electrochemical parameters are the maximum solid-phase lithium-ion concentrations in the positive and negative electrodes. Secondly, based on the incremental capacity-differential voltage (IC-DV) method, this chapter conducts the corresponding quantitative analysis of the LAM and the LLI, which are two lithium-ion degradation modes. Among them, LAM obtains quantification results through IC curves, and LLI obtains quantification results through DV curves. Finally, this chapter studies two correlation analysis methods based on multidimensional health indicators and health state, including Pearson correlation analysis (PCA) method and grey relational analysis (GRA) method, and explains the importance of strong correlation verification between health indicators and health state.

Keywords Cooperation competition particle swarm optimization · Important electrochemical parameters · Incremental capacity-differential voltage · Loss of active material · Loss of lithium inventory · Pearson correlation analysis · Grey relational analysis

As the input of health state neural network estimation model, the selection and extraction methods of health indicators are very important. Currently, common health state estimation models generally use parameters such as voltage, current, voltage drop rate, and cycle times as health indicators for model training and use. However, such methods are usually only applicable to one working condition. When working conditions change, the network trained by this kind of model cannot accurately describe the mathematical relationship between health indicators and health state. Therefore, such methods have the disadvantage of insufficient universality. The maximum solid-phase lithium-ion concentrations in the positive and negative electrodes in the

Q. Huang et al., *Long-Term Health State Estimation of Energy Storage Lithium-Ion Battery Packs*, https://doi.org/10.1007/978-981-99-5344-8_3

electrochemical model of energy storage lithium-ion batteries are directly related to the health state of the battery. It changes through internal reactions in the battery and is not affected by working conditions. In addition, two degradation modes, LLI and LAM, occur during the degradation process of energy storage lithium-ion batteries, which cannot be avoided and are closely related to health state. Significantly, they are not affected by the changes of working conditions. Therefore, this chapter selects the maximum solid-phase lithium-ion concentrations in the positive and negative electrodes and the two degradation modes mentioned above as health indicators for energy storage lithium-ion batteries. It is particularly noteworthy that the above two degradation modes can only be analyzed under low-rate constant current conditions for energy storage lithium-ion batteries. Therefore, under non-low-rate constant current conditions, only the maximum solid-phase lithium-ion concentrations in the positive and negative electrodes are extracted as the health indicators for energy storage lithium-ion batteries. It is worth noting that the quantized values of the degradation modes exist as characteristic feature values.

3.1 Research on Two-Parameter Identification of ESP Model Based on Improved Cooperative Competitive Particle Swarm Optimization Algorithm

3.1.1 Research on the Basic Composition of Improved Cooperative Competitive Particle Swarm Optimization Algorithm

Particle swarm optimization (PSO) algorithm searches for randomly initialized particles in space, and the advantages and disadvantages of the particles are calculated by the fitness value of the target function at the corresponding location [1]. These particles continuously adjust the speed and direction of movement in the search space based on their previous search experience and group search experience to iteratively calculate the fitness value of the objective function corresponding to each random solution at each discrete time, thereby evaluating the quality of the random solution and obtaining the location of the optimal particles, which is to obtain the optimal solution in the search space [2].

After obtaining the objective function, each particle in the PSO algorithm independently searches for the optimal solution in the search space. After the algorithm is iterated once, the individual optimal value of all particles is found as the global optimal solution of the population based on the fitness value. All particles in the population are updated based on the current individual optimal value and the population optimal value. The iterative updating process of the speed and position of each particle during each search process is shown in Eq. (3.1).

$$\begin{cases} v_{i,d}(k+1) = v_{i,d}(k) + c_1 r_1 \left(p_{best,i,d}(k) - x_{i,d}(k)\right) + c_2 r_2 \left(g_{best,d}(k) - x_{i,d}(k)\right) \\ x_{i,d}(k+1) = x_{i,d}(k) + v_{i,d}(k+1) \\ X_i = \left(x_{i,d}, x_{i,2}, \ldots x_{i,D}\right)^T, V_i = \left(v_{i,1}, v_{i,2}, \ldots v_{i,D}\right)^T \\ p_{best,i} = \left(p_{best,i,1}, p_{best,i,2}, \cdots p_{best,i,D}\right), g_{best} = \left(g_{best,1}, g_{best,2}, \cdots g_{best,D}\right) \end{cases} \tag{3.1}$$

In Eq. (3.1), for the particle i in the d-dimension, $v_{i,d}$, $x_{i,d}$ and $p_{best,i,d}$ represent the speed, the position and the current individual optimal solution, respectively. For the d-dimension, $g_{best,t,d}$ represents the current population optimal solution. For the particle i, V_i, X_i and $p_{best,t,i}$ represent the speed set, the position set and the current individual optimal solution, respectively. In addition, g_{best} represents the current population optimal solution, k represents the current number of iterations, r_1 and r_2 represent random numbers, c_1 and c_2 represent learning factors, and D represents the dimension of search space.

In addition, the first term on the right side of the d-dimensional speed vector component formula in Eq. (3.1) is called the memory term, reflecting the impact of the motion speed at the previous time on the current time. The second term on the right side of the formula is called the self-cognition term, reflecting the impact of the particle's previous search experience on its current motion speed. The third term on the right side of the formula is called the group cognition term, reflecting the cooperative relationship between particles. Under the joint action of the above three parts, a more accurate global optimal solution can be obtained at the end of the iteration. The general flow of the PSO algorithm is shown in Fig. 3.1.

In Fig. 3.1, for the particle i, $J_{p,present,i}$, $J_{p,best,i}$, $p_{present,i}$ and $p_{best,i}$ represent the present fitness, the fitness corresponding to the optimal identification results, the present position, and the position corresponding to optimal identification results. Additionally, g_{best} represents the optimal solution of the population, k represents the present number of iterations, and k_{iter} represents the maximum number of iterations.

However, the optimization results of the above basic PSO algorithm are not necessarily global optimal solutions. If the memory term in the d-dimensional speed vector component update formula for a particle has too much influence, it can be seen from Eq. (3.1) that the particle performs the linear search with a constant speed on the d-dimension, which may lead to missing the optimal solution. To address this shortcoming, a dynamically changing weight factor is introduced into the PSO algorithm, and the improved algorithm is called the standard PSO algorithm. Compared to the standard PSO algorithm, the speed update expression is updated as shown in Eq. (3.2).

$$v_i(k+1) = wv_i(k) + c_1 r_1 \left(p_{best,i} - x_i(k)\right) + c_2 r_2 (g_{best} - x_i(k)) \tag{3.2}$$

In Eq. (3.2), w is the weight factor whose value can affect the speed of particles at the next moment. A larger weight factor is conducive to global search, while a smaller weight factor is conducive to local search. At the beginning of the search, a larger weight factor can increase the probability of traversing the solution space,

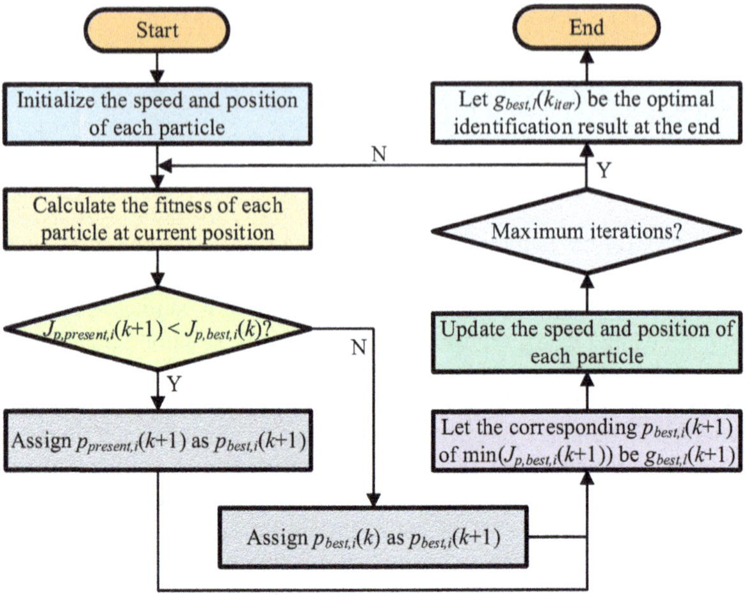

Fig. 3.1 The general flow of the PSO algorithm

and at the end of the search, a smaller weight factor can increase the probability of finding the global optimal solution. Therefore, setting appropriate weighting factors can improve the efficiency and accuracy of the algorithm in the optimization process. In this book, a linear decreasing function about the number of iterations is used to represent the dynamic change of the weight factor. The expression of the weight factor is shown in Eq. (3.3).

$$w = w_{\max} - \frac{w_{\max} - w_{\min}}{k_{iter}} k \tag{3.3}$$

In Eq. (3.3), w_{max} is the maximum value of the weight factor, w_{min} is the minimum value of the weight factor, k is the present number of iterations, and k_{iter} is the maximum number of iterations. From Eqs. (3.2) and (3.3), it can be seen that the memory term in the improved d-dimensional speed vector component update formula changes with w and is no longer affected by a fixed value. Therefore, standard PSO algorithms can enhance optimization efficiency and accuracy. However, the standard PSO algorithm has the problem of lacking evolutionary diversity of particle swarm optimization. As the name implies, insufficient evolutionary diversity of particle swarm can lead to a consistent evolutionary direction for the entire population. The result of this problem is that the standard PSO algorithm is desperate in the optimization process, making it difficult to jump out of a small search space, and it may occur that the local optimal solution is obtained due to premature algorithm. To address this deficiency, this book improves the standard PSO by introducing cooperation and competition strategies for proposing the CCPSO algorithm. These two strategies increase the diversity of particle swarm evolution by dividing population types and expanding particle search modes, respectively.

Cooperation strategies for optimizing standard PSO algorithms fall into two broad types. The first type is to coordinate each dimension in the optimization results so that each dimension develops in the optimal direction. This type of cooperation strategy does not consider optimal fitness, and is suitable for situations with multiple dimensions of optimization results. The second type is to divide particle swarm evolution regions. In such cooperation strategies, the particle swarm as a whole is divided into multiple relatively independent evolutionary regions, and information about the optimization results among regions can be exchanged and shared through a certain variable frequency. The goal of this type of cooperation strategy is to obtain optimization results corresponding to the optimal fitness with high efficiency, and it is suitable for situations where the optimization results have fewer dimensions. According to the needs of this book, in order to increase the diversity of particle swarm evolution in the standard PSO algorithm, this article introduces the second type of cooperation strategy based on the standard PSO.

Introducing the cooperation strategy into the standard PSO algorithm can divide the population into multiple subpopulations, which enables each subpopulation to undergo differential evolution in different ways. During each evolutionary process of the population, information is shared and exchanged among sub populations using an appropriate communication frequency function. In this cooperation strategy, communication between subpopulations needs to meet the communication frequency function shown in Eq. (3.4).

$$\begin{cases} f(t) = rand(0, 1) \\ F(t) = \dfrac{k}{k_{iter}} \\ f(t) < F(t) \end{cases} \tag{3.4}$$

In Eq. (3.4), $f(t)$ represents the random function in the range of 0 to 1, $F(t)$ represents the communication frequency function, k represents the number of iterations under this evolutionary process, and k_{iter} represents the maximum number of iterations. When communicating among subpopulations, the subpopulation with the best population optimal solution is called m, and the population optimal solution corresponding to the subpopulation m is called $g_{best,m}$. In addition, the population optimal solution values of other subpopulations are set to $g_{best,m}$. When there is no communication between subpopulations, each subpopulation performs independent evolutionary optimization according to the standard PSO algorithm. The selection of communication frequency functions greatly affects the performance of the cooperation strategy. A larger communication frequency function results in the optimal population solution of most subpopulations being $g_{best,m}$, which hinders the evolutionary diversity of particle populations. A smaller communication frequency function will reduce the probability of cooperation among subpopulations, thereby reducing the effectiveness of the cooperation strategy. In this book, the communication frequency function is selected as a function that linearly increases with the number of iterations, making the cooperation strategy more focused on global search of each subpopulation in the early stage of optimization, and more focused on local cooperative search among subpopulations in the late stage of optimization.

In addition, based on the cooperation strategy, this book also introduces a competition strategy to expand the particle search modes. Introducing a competition strategy into the standard PSO algorithm expands the evolutionary search direction of each particle, which enables each particle to evolve at two different speeds during each iteration. Based on the comparison of the target fitness values of two homologous sub particles differentiated from each particle, the sub particle with a smaller fitness value is regarded as the basis for the next evolution. Therefore, after the introduction of the competition strategy, the *d*-dimensional speed vector component formula and the *d*-dimensional position vector component formula in Eq. (3.1) are updated as shown in Eq. (3.5).

$$\begin{cases} v_{i,d,t}(k+1) = w_t v_{i,d}(k) + c_1 r_1 \left(p_{best,i,d}(k) - x_{i,d}(k)\right) + c_2 r_2 \left(g_{best,d}(k) - x_{i,d}(k)\right) \\ x_{i,d,t}(k+1) = x_{i,d,t}(k) + v_{i,d,t}(k+1) \\ v_{i,d}(k) = \left\{ v_{i,d,t}(k) \middle| \min J\left(x_{i,t}(k)\right) \right\} \\ x_{i,d}(k) = \left\{ x_{i,d,t}(k) \middle| \min J\left(x_{i,t}(k)\right) \right\} \end{cases} \tag{3.5}$$

In Eq. (3.5), for the sub particle t of the particle i in the d-dimension, $v_{i,d,t}$ and $x_{i,d,t}$ represent the speed and the position. In addition, w_t represents the weight factor of the sub particle t, $x_{i,t}$ represents the position of the sub particle t of the particle i, k represents the number of iterations, and J represents the objective function.

In order to highlight the diversity of population evolution, in the competition strategy, the two evolution speeds of homologous sub particles include fast movement and slow movement. Rapidly moving sub particles have larger weight factors and are more focused on global search for optimal solutions, while slowly moving sub particles have smaller weight factors and are more focused on local search for optimal solutions. It is worth noting that if the competition strategy is combined with the proposed cooperation strategy, the population optimal solution in Eq. (3.5) no longer represents the overall population optimal solution, but rather the subpopulation population optimal solution.

Combining the standard PSO algorithm, the cooperation strategy and the competitive strategy, the basic process of the CCPSO algorithm can be summarized as follows:

(1) The total number N of particles in the search area, the maximum number k_{iter} of iterations of the algorithm, subpopulation category M, learning factors c_1 and c_2, the position $x_{i,d}$ of each particle in the d-dimension, and the running speed $v_{i,d,t}$ of each particle's subparticle t in the d-dimension are randomly initialized.

The initial position of each particle is assigned to the present individual optimal solution of the particle.

(2) The fitness values corresponding to the two homologous sub particles of each particle under this iteration based on the objective function are calculated, the smaller fitness value is retained, and the sub particle corresponding to the fitness value is regarded as the basis for the next evolution.

(3) For each particle, the fitness value of this iteration and the fitness value corresponding to the optimal solution of the previous iteration are compared, and the location coordinate with the smaller fitness value is assigned to the individual optimal solution of this iteration. For each subpopulation l, the optimal value of all individual optimal solutions of subpopulation l is taken as the population optimal solution $g_{best,l}$ for this iteration.

(4) The random function $f(t)$ and the communication frequency function $F(t)$ are compared. If $f(t)$ is smaller than $F(t)$ at this iteration number, the subpopulation with the best population optimal solution is selected as subpopulation m. The population optimal solution of subpopulation m is set to $g_{best,m}$, and $g_{best,m}$ is assigned to the population optimal solution of other subpopulation $g_{best,l}$. Otherwise, the population optimal solution $g_{best,l}$ of each subpopulation is independently calculated by each subpopulation.

(5) The speed, position, individual optimal solution, and group optimal solution of particles are updated according to Eq. (3.5). Significantly, if the updated parameter exceeds the preset range, the corresponding parameter value will be set to the preset limit value.

(6) Step (2) to step (5) are performed and cycled until the iteration is complete. The optimal solution of the subpopulation with the lowest fitness value is assigned to the global optimal solution of the entire population.

In the above CCPSO algorithm steps, the dynamic weighting strategy is applied, with the competition strategy reflected in step (2) and the cooperation strategy reflected in step (4). During each iteration, the competition strategy always precedes the cooperation strategy because the competition strategy is targeted at particles, while the cooperation strategy is targeted at subpopulations.

3.1.2 Research on Electrochemical Parameter Identification Process Based on Improved CCPSO Algorithm

The ESP model for energy storage lithium-ion batteries involves many electrochemical parameters, and some of them cannot be directly measured through experiments. For this book, the maximum solid-phase lithium-ion concentration in the positive and negative electrodes are the two most important electrochemical parameters, which directly affect the accuracy of the ESP model and the health state of the energy storage lithium-ion battery. In this book, parameter identification refers to comparing the voltage output of the ESP model with the actual voltage output data, and obtaining

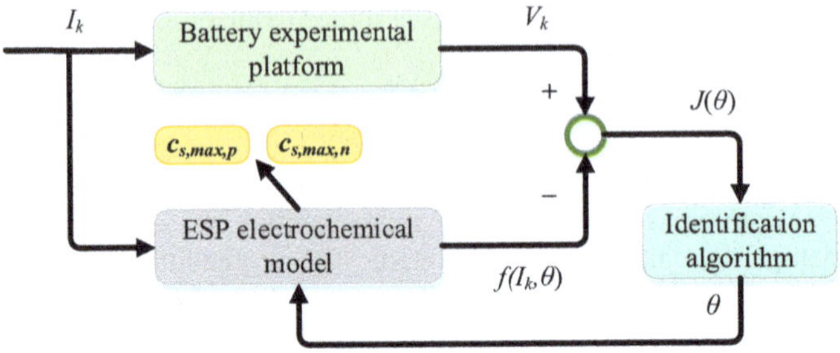

Fig. 3.2 The parameter identification structure diagram of the battery model

the optimal value of the parameter through the objective function. The core idea of the parameter identification algorithm is to minimize the error variance between the measured terminal voltage and the simulated terminal voltage of the ESP model. Therefore, the target function for parameter identification is shown in Eq. (3.6).

$$\min J(\theta) = \min \sum_{k=1}^{n} [V_k - f(I_k, \theta)]^2, \theta = \left(c_{s,\max,p}, c_{s,\max,n}\right) \tag{3.6}$$

In Eq. (3.6), k, V, I, f, θ and n denote the sampling time, the measured terminal voltage, the operating current, the simulated terminal voltage, the identified parameter set, and the maximum sampling time, respectively. Combining the above objective function, the parameter identification structure diagram of the battery model is shown in Fig. 3.2.

As the CCPSO algorithm steps mentioned in the previous content are only a general process, for the purpose of this book, the CCPSO algorithm requires the identification of two important electrochemical parameters based on actual terminal voltage data. Therefore, the specific flowchart for identifying the maximum solid-phase lithium-ion concentration in the positive and negative electrodes in the ESP model using the CCPSO algorithm is shown in Fig. 3.3.

In Fig. 3.3, for the subpopulation l, $J_{g,best,l}$ and $g_{best,l}$ represent the fitness value of the optimal identification results and the optimal identification results, respectively.

3.2 Quantification of Two Degradation Modes Based on Incremental Capacity-Differential Voltage Method

The degradation modes are a set of factors that cause battery degradation or a set of forms that exhibit battery degradation behavior. The degradation modes of lithium-ion batteries mainly include LLI, LAM and conductivity loss (CL).

Active lithium ions are lithium ions that complete the process of embedding and disengaging electrodes during normal battery operation. LLI is characterized

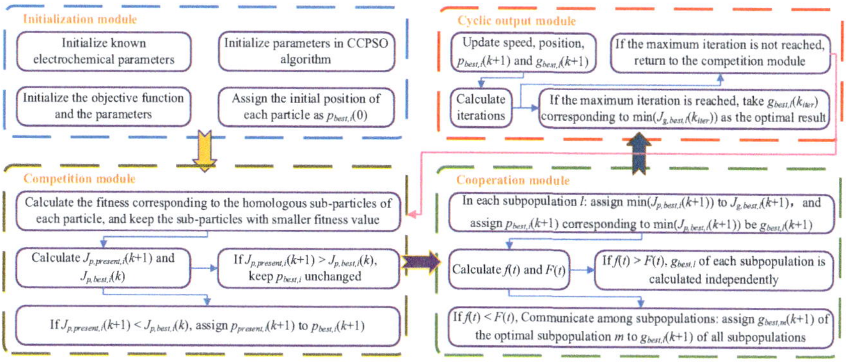

Fig. 3.3 The flowchart of parameter identification based on CCPSO algorithm

by a characteristic of battery capacity degradation. The active material serves as a carrier for lithium ions embedded in the electrodes during the reaction process of the battery. Therefore, LAM can also indirectly lead to the degradation of battery capacity. CL mainly characterizes the power degradation characteristics of lithium-ion batteries, so it is not mentioned in this book. During battery degradation process, battery degradation occurs due to LAM, LLI, and other reasons. The same battery undergoes constant current charging and discharging, and electrochemical reactions occur differently within the battery with different degradation stages.

Since the IC-DV method introduced in this book is used to quantify the two above-mentioned degradation modes, changes in the IC-DV curves at different rate and temperature are not considered. Thence, only changes in the IC-DV curves at different degradation stages are considered. Under different degradation stages, the constant current open-circuit voltage curve has a little difference, making it difficult to visually distinguish the differences in open-circuit voltage curves with different degradation stages. The IC-DV method can convert the flat phase of the curve into the $\mathrm{d}Q/\mathrm{d}U$ of the IC curve and the $\mathrm{d}U/\mathrm{d}Q$ of the DV curve. By observing the changes in IC curves and DV curves, the characteristics of IC-DV curves of batteries with different degradation stages is analyzed, thereby quantifying the battery degradation modes, and laying the foundation for subsequent battery health state estimation.

3.2.1 Research on LAM Quantification Based on the IC Curve

The IC method is a method for analyzing battery degradation characteristics based on IC curves. In practical applications, the ratio of capacity change ΔQ and voltage step ΔU is defined as the IC, so the expression for the IC method is shown in Eq. (3.7).

$$IC = \frac{\Delta Q}{\Delta U} \tag{3.7}$$

It can be seen from Eq. (3.7) that for obtaining the IC curve, the support of a capacity-voltage curve is required. Therefore, it is necessary to conduct low-rate charging or discharging experiments on the battery. The reason why high-rate current is not used is that under high-rate conditions, the intercalate and deintercalate processes of lithium ions are hindered, resulting in a decrease in the maximum available capacity of the battery. The schematic diagram of IC curve with voltage as abscissa and IC as ordinate is shown in Fig. 3.4.

As shown in Fig. 3.4, under the constant ΔU, the smaller the ΔQ, the smaller the corresponding IC value, indicating that the capacity changes slowly within this voltage range. On the contrary, the larger the ΔQ, the larger the corresponding IC value, indicating that the capacity changes quickly within this voltage range. Among them, the peaks in the IC curve represent the most significant incremental changes in capacity, and they can to some extent characterize the structural characteristics and material changes within the lithium-ion battery. In addition, the peak point of the IC curve is located in the plateau region of the battery discharge curve, which can characterize the sensitivity of the battery to capacity changes during stable discharge. Carlos Pastor Fernández [3] and David Anseán [4] pointed out that LAM is mainly reflected in the change of the left peak of the IC curve. Therefore, this book constructs a quantification method for LAM based on IC curves, and the expression of LAM quantification is shown in Eq. (3.8).

$$LAM = \left| \frac{\frac{\Delta Q_{IC}}{\Delta U_{IC}} - \frac{\Delta Q_{IC}}{\Delta U_{IC}}|1}{\frac{\Delta Q_{IC}}{\Delta U_{IC}}|1} \right| \tag{3.8}$$

In Eq. (3.8), $\Delta Q_{IC}/\Delta U_{IC}$ and $\Delta Q_{IC}/\Delta U_{IC}|1$ are the present peak value and the initial peak value of the most obvious peak in the left side of the IC curve, respectively.

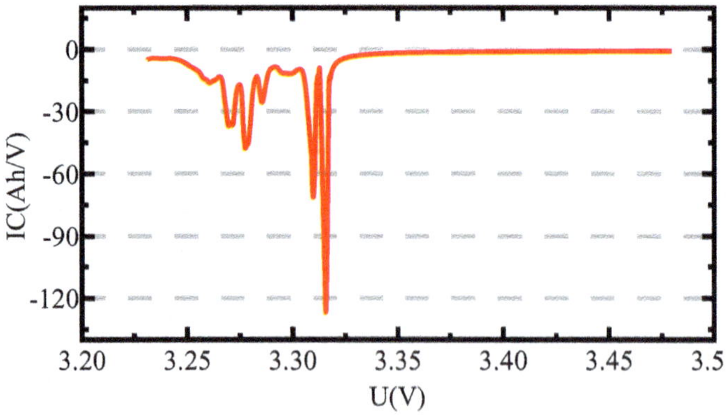

Fig. 3.4 The schematic diagram of IC curve

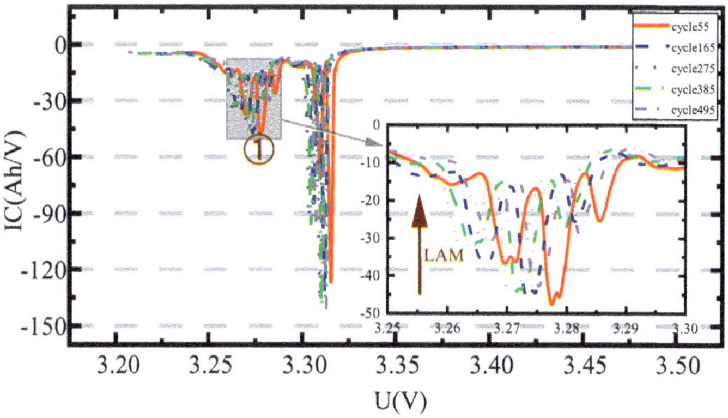

Fig. 3.5 The quantification schematic diagram of LAM

The LAM quantification method based on IC curves is able to analyze the degradation characteristics and LAM change trends of batteries from the time domain perspective. Figure 3.5 is the quantification schematic diagram of LAM.

Figure 3.5 shows the quantification diagram of LAM in IC curves, and the peak 1 marked in Fig. 3.5 is the peak with the most significant change on the left side. In addition, from the enlarged schematic diagram in Fig. 3.5, it can be seen that with the increase of degradation cycles, both the IC value and the LAM value at peak 1 demonstrate an upward trend.

3.2.2 Research on LLI Quantification Based on the DV Curve

The DV method is a method for analyzing battery degradation characteristics based on DV curves. In practical applications, the ratio of capacity change ΔU and voltage step ΔQ is defined as the DV, so the expression for the DV method is shown in Eq. (3.9).

$$DV = \frac{\Delta U}{\Delta Q} \tag{3.9}$$

Similar to the IC method, it is also necessary to conduct low-rate charging or discharging experiments on the battery to obtain DV curves. The schematic diagram of DV curve with capacity as abscissa and DV as ordinate is shown in Fig. 3.6.

As shown in Fig. 3.6, under the constant ΔQ, the smaller the ΔU, the smaller the corresponding DV value, indicating that the voltage changes slowly within this capacity range. On the contrary, the larger the ΔU, the larger the corresponding DV value, indicating that the voltage changes quickly within this capacity range. The value and the location of the peaks in the DV curve can characterize the internal changes in the lithium-ion battery, reflecting the sensitivity of the battery to voltage

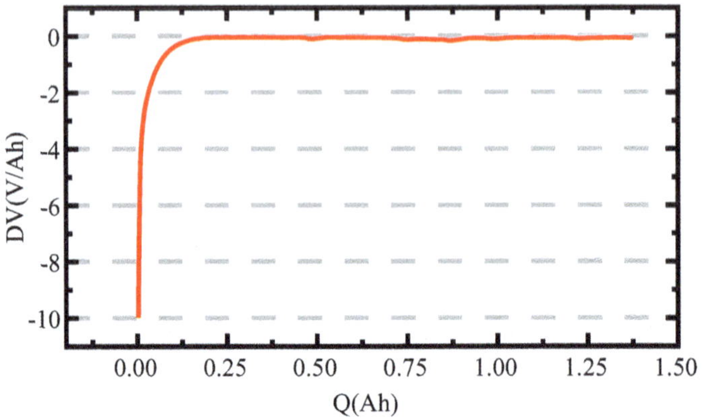

Fig. 3.6 The schematic diagram of DV curve

changes. From the above description, it can be seen that DV and IC are two opposite concepts. Meinert Lewerenz [5] believed that LLI can be reflected to some extent by the offset of the DV curve along the coordinate axis. Therefore, this book establishes a LLI quantification method based on DV curves, and the expression of LLI quantification is shown in Eq. (3.10).

$$LLI = \left| \frac{Q_{DV} - Q_{DV,1}}{Q_{DV,1}} \right| \tag{3.10}$$

In Eq. (3.10), Q_{DV} and $Q_{DV,1}$ are the current capacity and the initial capacity of the most obvious peak in the DV curve, respectively. The LLI quantification method based on DV curves is able to analyze the degradation characteristics and LLI change trends of batteries from the time domain perspective. Figure 3.7 is the quantification schematic diagram of LLI.

Figure 3.7 is the quantification diagram of LLI in DV curves. From the enlarged diagram in Fig. 3.7, it can be seen that with the increase of degradation cycles, the DV value and the LLI value at the most significant peak in DV curves show a downward trend and a left shift trend, respectively.

3.3 Correlation Analysis of Health Indicators

When using a data-driven model to estimate the health state of lithium-ion batteries, the selection of appropriate health indicators is the primary issue. In order to improve the training efficiency of the data-driven model and the accuracy of health state estimation results, it is necessary to adopt health indicators closely related to health state degradation as inputs of the data-driven model. The higher the correlation between the adopted health indicators and health state, the more accurate the training

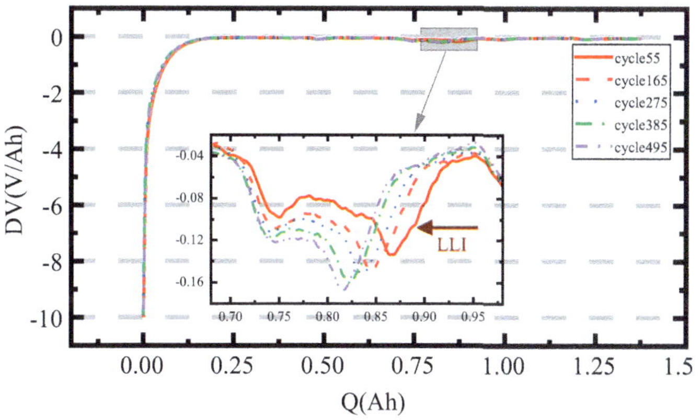

Fig. 3.7 The quantification schematic diagram of LLI

results will be [6]. In this book, PCA method and GRA method are used to analyze whether the maximum solid-phase lithium-ion concentration in the positive electrode, the maximum solid-phase lithium-ion concentration in the negative electrode, the LLI, and the LAM are suitable as input parameters for the data-driven health state estimation model.

3.3.1 Pearson Correlation Analysis Based on Multidimensional Health Indicators and Health State

The PCA method is widely used to detect the degree of linear correlation between two continuous variables. Its core idea is to calculate the Pearson correlation coefficient between variables. Pearson correlation coefficient is used to quantitatively analyze the linear correlation between variables, and its value is the quotient of covariance and standard deviation between variables. The specific calculation process is shown in Eq. (3.11).

$$\begin{aligned}\rho_{X,Y} &= \frac{\operatorname{cov}(X,Y)}{\sigma_X\sigma_Y} = \frac{E((X-\mu_X)(Y-\mu_Y))}{\sigma_X\sigma_Y} \\ &= \frac{E(XY)-E(X)E(Y)}{\sqrt{E\left(X^2\right)-E^2(X)}\sqrt{E\left(Y^2\right)-E^2(Y)}}\end{aligned} \tag{3.11}$$

In Eq. (3.11), X and Y represent sequences for analysis, ρ represents Pearson correlation coefficient, *cov* represents covariance, μ represents the arithmetic mean function, σ represents standard deviation, and E represents mathematical expectation. In this book, X represents the sequence of changes in health state of the energy storage lithium-ion battery with the increase of degradation cycles, and Y represents the sequence of changes in the maximum solid-phase lithium-ion concentrations or degradation modes with the increase of degradation cycles. Although PCA has the advantages of fast computation speed, Pearson correlation coefficients lack the expression of nonlinear relationships.

3.3.2 Grey Relational Analysis Based on Multidimensional Health Indicators and Health State

Due to the fact that the PCA method can only represent linear correlations between sequences, it is not sensitive to nonlinear relationships. Therefore, this section introduces the GRA method. The core idea of GRA method is to analyze the curve similarity between sequences to measure the degree of correlation between sequences, which is a supplement to the shortcomings of Pearson correlation coefficient. The degree of correlation between changes in two sequences is called the degree of correlation, which is positively correlated with the trend of changes in the sequence. The GRA method consists of four steps:

(1) Determine the reference sequence and the comparison sequences. In this book, the reference sequence is the sequence of changes in health state of energy storage lithium-ion batteries with increasing degradation cycles, while the comparison sequences are the sequences of changes in the maximum solid-phase lithium-ion concentrations in the positive electrode and the negative electrode, LAM, and LLI with increasing degradation cycles. The expressions for the reference sequence and comparison sequence are shown in Eq. (3.12).

$$\begin{cases} X_i = \{x_i(k)|k = 1, 2, ..., n\} \\ Y = \{y(k)|k = 1, 2, ..., n\} \end{cases} \tag{3.12}$$

In Eq. (3.12), X and Y represent the comparison sequence and reference sequence, respectively, i represents the sequence number of the comparison sequence, k represents the sequence number of each point in the sequence, and n represents the sequence length.

(2) Pre-process the reference sequence and the comparison sequences. Different data dimensions can have a significant impact on the analysis results of the GRA method, so it is necessary to perform dimensionless processing on the data in the sequences. This book chooses the mean processing method to pre-process the sequences, and the expression of the mean processing method is shown in Eq. (3.13).

$$x_{trans} = \frac{x}{\text{mean}(X)} \tag{3.13}$$

In Eq. (3.13), x represents a specific value in the sequence, x_{trans} represents the dimensionless value of x.

(3) Calculate the relational coefficient between the reference sequence and the comparison sequences. Therefore, the difference between curves is used as a quantitative indicator of the degree of correlation, and the calculation equation for the relational coefficient is shown in Eq. (3.14).

$$\begin{cases} \xi_i(k) = \dfrac{\min\limits_i \min\limits_k \Delta_i(k) + \rho \min\limits_i \min\limits_k \Delta_i(k)}{\Delta_i(k) + \rho \min\limits_i \min\limits_k \Delta_i(k)} \\ \Delta_i(k) = |y(k) - x_i(k)| \end{cases} \tag{3.14}$$

In Eq. (3.14), i represents the sequence number in a comparison sequence, k represents the sequence number for each point in this sequence, ξ represents the relational coefficient between each comparison sequence and the reference sequence at each point, and ρ represents the resolution coefficient.

(4) Calculate the relational degree between the reference sequence and the comparison sequences. The correlation coefficients can only describe the degree of correlation between the comparison sequences and the reference sequence at each point, and cannot be used for overall similarity comparison between sequences. Therefore, the concept of relational degree is proposed, which is the mean value of the relational coefficients under the corresponding comparison sequence. The calculation equation for relational degree is shown in Eq. (3.15).

$$r_i = \frac{1}{n}\sum_{i=1}^{n} \xi_i(k), \ (k = 1, 2, \ldots, n) \tag{3.15}$$

In Eq. (3.15), n represents the length of the sequence, r represents the grey relational value, which is between 0 and 1 and is positively correlated with the relation between sequences.

3.4 Summary

This chapter delves into the extraction of health indicators for multidimensional energy storage lithium-ion batteries. For the health state estimation topic of this book, the input of the data-driven model is referred to as health indicators. Appropriate and universal health indicators are crucial for health state estimation results. The CCPSO algorithm is used to identify the maximum solid-phase lithium-ion concentrations in the positive and negative electrodes of the ESP cell model. Two degradation modes related the capacity are quantified by using the IC-DV method, and the above parameter sequences are regarded as health indicators. The identification results are verified by the model accuracy, and significantly, the quantification sequences of degradation modes are only applicable to low-rate constant current conditions.

References

1. Zhuoyan Wu, L.Y., Ran Xiong, Shunli Wang, Wei Xiao, Yi Liu, Jun Jia and Yanchao Liu. 2022. A novel state of health estimation of lithium-ion battery energy storage system based on linear decreasing weight-particle swarm optimization algorithm and incremental capacity-differential voltage method. *International Journal of Electrochemical Science* 17(7): 1–32.
2. Qin, T., S. Zeng, and J. Guo. 2015. Robust prognostics for state of health estimation of lithium-ion batteries based on an improved PSO–SVR model. *Microelectronics Reliability* 55 (9–10): 1280–1284.
3. Pastor-Fernández, C., et al. 2017. A comparison between electrochemical impedance spectroscopy and incremental capacity-differential voltage as Li-ion diagnostic techniques to identify and quantify the effects of degradation modes within battery management systems. *Journal of Power Sources* 360 (31): 301–308.
4. David, A., et al. 2019. Lithium-ion battery degradation indicators via incremental capacity analysis. *IEEE Transactions on Industry Applications* 55 (3): 1–10.
5. Lewerenz, M., et al. 2017. Differential voltage analysis as a tool for analyzing inhomogeneous aging: A case study for LiFePO4 |Graphite cylindrical cells. *Journal of Power Sources* 368 (15): 57–67.
6. Kaiquan, L., W. Yujie, and C. Zonghai. 2022. A comparative study of battery state-of-health estimation based on empirical mode decomposition and neural network. *Journal of Energy Storage* 54 (10): 105333–105348.

Chapter 4
Research on Health State Estimation Method of the Lithium-Ion Battery Pack

Abstract The purpose of this chapter is to establish a neural network model which is suitable for cell health state estimation and calculate the overall health state of the energy storage lithium-ion battery pack based on cell health state. This chapter firstly elaborates on the improvement principle of the nonlinear coefficient temperature decreasing step size simulated annealing (NSA) algorithm, with a focus on the NSA strategy. On this basis, an improved nonlinear coefficient temperature decreasing step size simulated annealing-back propagation (NSA-BP) neural network model using the NSA algorithm to optimize the initial parameters of the back propagation (BP) neural network is proposed. Secondly, this chapter investigates the health state estimation process based on the improved NSA-BP model. The definition methods of cell health state and battery pack health state are introduced in detail. Finally, this chapter combines cell model, pack model, cell model parameter identification algorithm, cell quantification method of degradation modes, correlation analysis methods, and neural network algorithm to construct a long-term health state estimation model.

Keywords Nonlinear coefficient temperature decreasing step size · Simulated annealing · Back propagation · Neural network model · Health state estimation model · Battery pack health state

4.1 Research on the Basic Composition of Improved NSA-BP Model

4.1.1 Research on the Improvement of NSA Algorithm

The basic idea of simulated annealing (SA) is derived from the annealing process of solid materials [1]. The state when the simulation reaches the minimum energy is the optimal solution of the system objective function. This algorithm adds escape probability to introduce inferior solutions during the search process, so that the search results can avoid local optima as much as possible and converge to the global optimal solution. The steps of the SA algorithm can be summarized as follows:

Q. Huang et al., *Long-Term Health State Estimation of Energy Storage Lithium-Ion Battery Packs*, https://doi.org/10.1007/978-981-99-5344-8_4

(1) Randomly assign the corresponding values to the initial temperature T_0, initial solution S_0, Markov chain length L and total number of iterations K_{max}, and consider that the initial solution S_0 is the initial optimal solution.
(2) Set the system objective function to C, randomly select a new solution S_{new} within the nearest subset of the current optimal solution S, and calculate the objective function $C(S_{new})$ of the new solution.
(3) Calculate the increment of the objective function ΔC. The calculation expression is shown in Eq. (4.1).

$$\Delta C = C(S_{new}) - C(S) \tag{4.1}$$

(4) According to the Metropolis criterion, the probability P accepts S_{new} as the optimal solution for the current temperature system, otherwise the original S is maintained as the optimal solution. The calculation formula for probability P is shown in Eq. (4.2).

$$P = \begin{cases} 1\,, \Delta C < 0 \\ e^{-\frac{\Delta C}{T}}\,, \Delta C > 0 \end{cases} \tag{4.2}$$

(5) Perform iterations at the current temperature T and cycle through steps (2) to (4).
(6) Make the current temperature T decrease at a certain speed. In the classic SA algorithm, the temperature is generally set to exponentially decrease, and the specific expression is shown in Eq. (4.3).

$$T_{new} = \lambda T\,, 0 < \lambda < 1 \tag{4.3}$$

In Eq. (4.3), λ represents the temperature drop coefficient, T represents the temperature before the update, and T_{new} is the new temperature value after one update.
(7) After updating the temperature, perform the external cycle from steps (2) to (6) until the temperature drops to the set termination temperature T_{final}, and output the current solution as the optimal solution for the system.

In the steps of the SA algorithm mentioned above, the initial temperature value T_0 and the final temperature value T_{final} need to be assigned based on specific practical problems. The process of exponential annealing is neither fast nor slow, and it is the most commonly used cooling function in the classic SA algorithm. The schematic diagram of exponential annealing strategy is shown in Fig. 4.1.

As shown in Fig. 4.1, the exponential annealing criterion exhibits a large rate of change in the initial stage of temperature decrease, which is not conducive to the algorithm seeking the optimal solution at a higher temperature in the initial stage, and may lead to the optimization results falling into local optimum. Therefore, this book introduces and proposes the NSA strategy to overcome the above problems as much as possible. In the NSA algorithm, the exponential annealing criterion is replaced

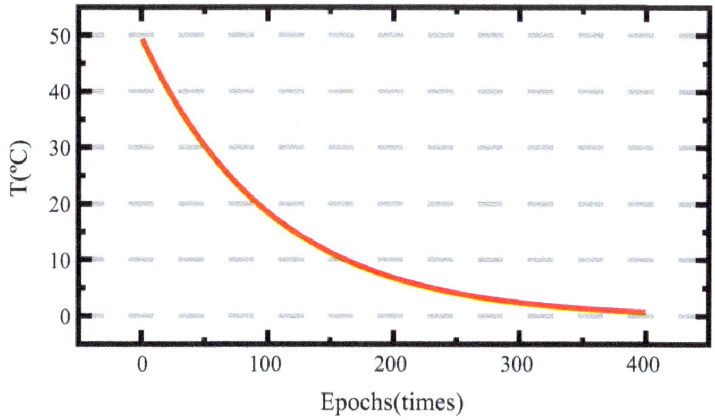

Fig. 4.1 Schematic diagram of exponential annealing

with a novel nonlinear coefficient temperature decreasing step annealing criterion. The improved method can improve the search efficiency of the algorithm and avoid local optimum as much as possible. The expression for temperature changes in the improved NSA algorithm is shown in Eq. (4.4).

$$T = \frac{T_0 - T_{final}}{2} \cos \frac{\pi}{K_{\max}} K + \frac{T_0 + T_{final}}{2} \tag{4.4}$$

In Eq. (4.4), T represents the updated temperature, T_0 represents the initial temperature, T_{final} represents the final temperature, K_{max} represents the maximum number of iterations in the NSA algorithm, and K represents the current number of iterations. The schematic diagram of the NSA strategy is shown in Fig. 4.2.

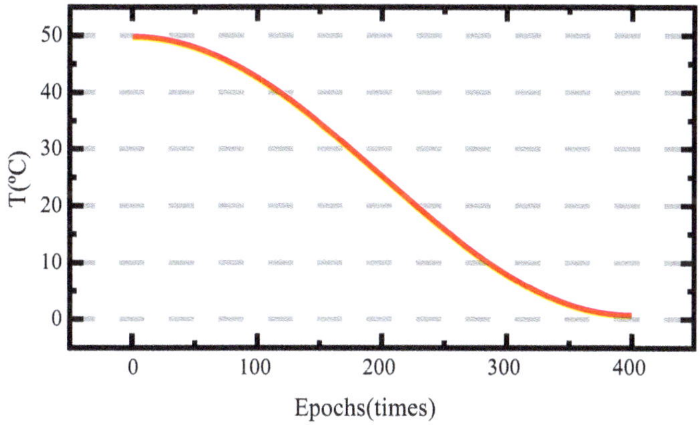

Fig. 4.2 Schematic diagram of NSA strategy

From Fig. 4.2, it can be seen that the NSA function has a slow temperature decrease rate at the initial stage, which is conducive to exploring the optimal solution at a larger temperature in the early stage, and a slow temperature decrease rate at the end, which is conducive to conducting a small range of accurate searches at a smaller temperature change rate in the later stage. Compared with the calculation steps of the SA algorithm, the proposed improved NSA algorithm only needs to improve the exponential annealing criterion Eq. (4.3) in step (6) of the SA algorithm to the nonlinear coefficient temperature decreasing step size cooling criterion Eq. (4.4).

4.1.2 Research on the Optimization of BP Neural Network Based on NSA Algorithm

From the perspective of bionics, the neural network model simulates the structure and characteristics of the animal brain, and has a good effect on the description of nonlinear systems such as lithium-ion batteries [2]. Neural networks originate from the nervous system of animals and have advantages such as high fault tolerance, strong adaptability, strong anti-interference ability, and strong self-learning ability. Neural networks do not require an understanding of the system [3, 4]. They can automatically obtain system features during the training process and simulate the nonlinear mapping relationship between system inputs and outputs through learning. The parallel structure of neural networks enables each neuron to independently process the received information, greatly improving its running speed [5, 6]. BP neural network is a simple and practical neural network model, so it is often used to solve the online estimation of nonlinear system problems [7, 8]. Because the essence of the battery is electrochemical reaction, the battery is a highly complex nonlinear system [9]. This book uses a neural network that takes the maximum solid-phase lithium-ion concentrations in the positive and negative electrodes, LLI, and LAM as inputs to map the battery and estimate the battery health state.

The characteristics of the BP neural network is mainly determined by its structure and learning rule. The BP neural network is structurally a kind of forward feedback neural network, in which signals and errors propagate forward and backward, respectively. Among them, the errors of backpropagation are used to adjust the weights between neurons to achieve the expected output value after multiple iterations. The network structure diagram adopted in this book is shown in Fig. 4.3.

From Fig. 4.3, it can be seen that the BP neural network model includes an input layer, a hidden layer and an output layer. Neurons are the basic structure in neural network models, which contain information about connection chains, adders and transfer functions. The states of neurons largely determine the output value. The input and output relationship between adjacent layers of the BP network is shown in Eq. (4.5).

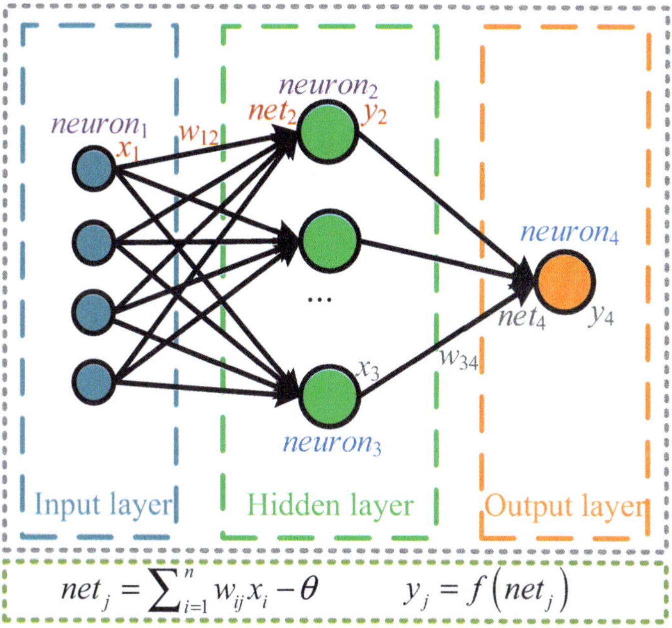

Fig. 4.3 The network structure diagram of adopted BP model

$$\begin{cases} net_j = \sum_{i=1}^{n} w_{ij} x_i - \theta \\ y_j = f(net_j) \end{cases} \tag{4.5}$$

In Eq. (4.5), x_i and w_{ij} represent the signal from neuron i and the weight between neuron i and neuron j, respectively. In addition, n represents the number of neurons input to neuron j, θ represents the threshold of neuron j, net_j represents the input of neuron j, $f()$ represents the transfer function, and y_j represents the output of neuron j. The square error function is used as the objective function, as shown in Eq. (4.6).

$$E = \frac{1}{2} \sum_{k=1}^{N} \left(y_{t,k} - y_{BP,K} \right)^2 \tag{4.6}$$

Among them, E represents the total error output by the system, k represents the sample number, and N represents the number of neurons in the output layer. When i and j in Eq. (4.6) represent neurons in the hidden layer and output layer, respectively, $y_{t,k}$ equals y_j. In the process of backpropagation, the errors of forward propagation can be used by the learning rule to update the weights of each layer until the errors reach the threshold. The learning rule is the updating rule of weights among neurons, including fixed memory learning methods, unsupervised learning methods and supervised learning methods. The connection weights in fixed memory learning methods are invariant, therefore the corresponding model accuracy is low. The connection weights in the unsupervised learning method can be adjusted independently, but are not affected by the feedback signal. In the supervised learning method, an evaluation standard will be set as the learning rule, and the connection weight value will be adjusted through feedback information to improve the accuracy of the model. Common learning rules include gradient descent method, quasi-Newton method and Levenberg–Marquardt (L-M) method. The calculation equation for the corresponding weight change varies with different learning rules. The training process flowchart of the BP neural network is shown in Fig. 4.4.

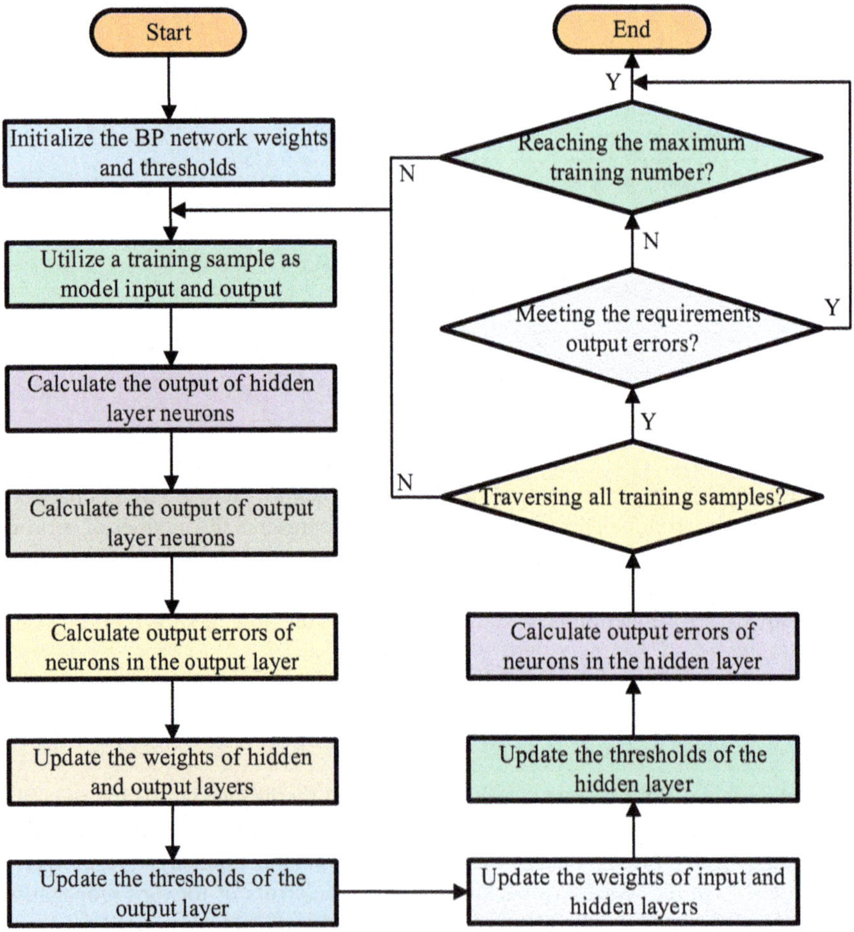

Fig. 4.4 The training process flowchart of the BP neural network

From Fig. 4.4, it can be seen that after all the training samples are taken out, the output error and maximum training frequency of the BP neural network determine whether the network stops or not. In addition, Fig. 4.4 also shows that feedback information can help the BP neural network continuously update connection weights and thresholds. Due to the different initial network parameters which are randomly generated by the BP neural network will have different impacts on the training of the entire network, this book introduces the NSA algorithm described in the previous section to search for the optimal initial parameters. The main function of the NSA algorithm in the NSA-BP model is to optimize the initial parameters of the BP neural network, and use the optimal initial parameters obtained after iterative updates as a new starting point for BP neural network training. Specifically, the flowchart of the NSA-BP neural network algorithm is shown in Fig. 4.5.

From Fig. 4.5, it can be seen that the main process of the NSA-BP model includes the following steps:

(1) Initialize the network structure and weight parameters of the BP model.
(2) Take the initial parameters of the BP neural network model as the target to be optimized in the NSA algorithm, and optimize the initial parameters of the BP neural network model.
(3) Input the optimal network parameters obtained through NSA algorithm optimization into the BP model.
(4) Train and test the NSA-BP neural network model.

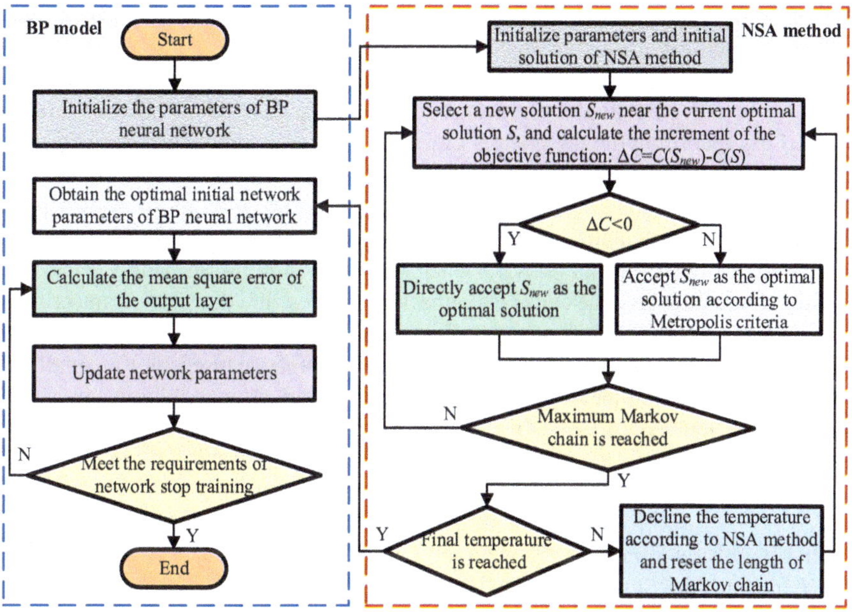

Fig. 4.5 The flowchart of the NSA-BP neural network algorithm

4.2 Research on Health State Estimation Strategy Based on Health Indicators and Improved NSA-BP Model

4.2.1 Health State Estimation Process Analysis Based on Improved NSA-BP Model

To study the health state estimation method based on the improved NSA-BP model, it is firstly necessary to understand the definition method of health state for energy storage lithium-ion batteries. In Chap. 1, it is mentioned that the capacity definition method of health state is more suitable for new energy storage systems. Therefore, this book uses this method to calculate health state, and the corresponding calculation equation is shown in Eq. (4.7).

$$SOH = \frac{Q}{Q_{rate}} \tag{4.7}$$

In Eq. (4.7), Q represents the current capacity of the energy storage lithium-ion battery, and Q_{rate} represents the rated capacity of the energy storage lithium-ion battery. Therefore, after obtaining the rated capacity of the battery and the capacity under a certain degradation cycle, the health state under this cycle is obtained. Due to the fact that in the NSA-BP neural network model, the health state of each cell is used as the output of the model. Therefore, after calculating the health state of each cell under different degradation cycles, the health state degradation sequence of each cell is obtained. By using this sequence as the output of the NSA-BP neural network model and using the parameter sequence strongly correlated with health state as the input of the NSA-BP model, the weights and other parameters of the network are trained. By recording the parameters of the network and conducting model testing, the health state of energy storage lithium-ion battery cells in other states is obtained. In addition, as the target of this book is the energy storage lithium-ion battery pack, it is necessary to calculate the pack health state after obtaining the cell health state. The model adopted in this book for energy storage lithium-ion battery packs is a multi-cell model where the cells are connected in series, and the series connection of the cells only increases the overall voltage without increasing the overall capacity. Therefore, in order to avoid overall performance defects caused by too small cell capacity, this book defines the health state of the energy storage lithium-ion battery pack as shown in Eq. (4.8).

$$SOH_p = \min(SOH_1, SOH_2, \ldots, SOH_n) \tag{4.8}$$

In Eq. (4.8), n represents the number of cells in the energy storage battery pack, SOH_n denotes the SOH of the n-th cell, and SOH_p signifies the overall pack SOH. Through the above-mentioned analysis, a flowchart for estimating health state of the energy storage lithium-ion battery pack based on the improved NSA-BP model is obtained, as shown in Fig. 4.6.

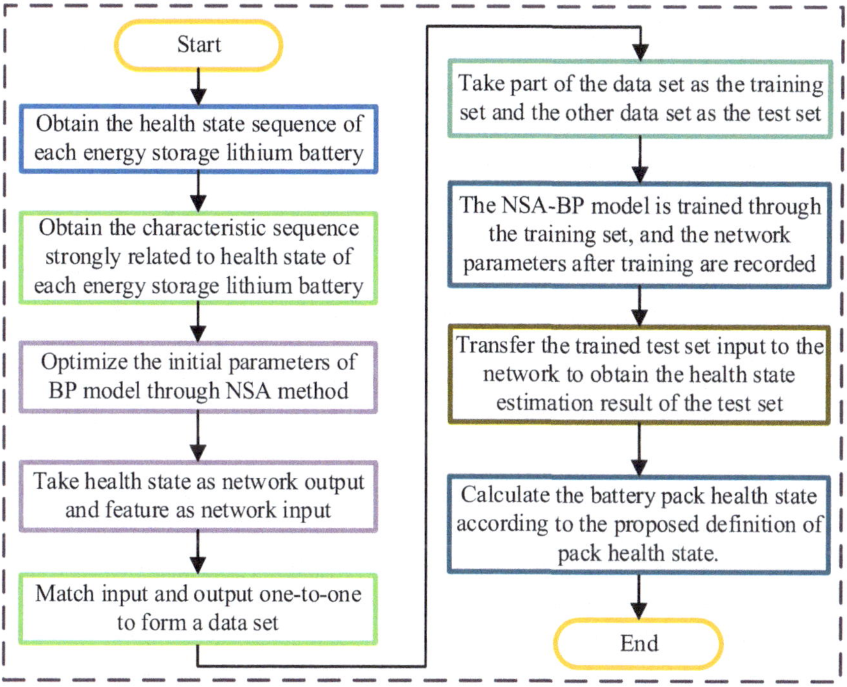

Fig. 4.6 The flowchart for estimating pack health state based on the improved NSA-BP model

As shown in Fig. 4.6, in the process of estimating health state of the energy storage lithium-ion battery pack, firstly, the health state and the parameter sequence strongly related to health state as the output and input of the estimation model are required for health state estimation. Among them, the input and output are matched one-on-one to form a dataset that includes training and testing sets. Secondly, there is a need to optimize the initial parameters of the BP neural network through the NSA algorithm to enhance the estimation performance of the neural network model. Then, it is necessary to train the network of the NSA-BP model through the training set data and record the network parameters after training. Finally, it is also necessary to transfer the input data from the test set to the trained network to obtain the health state estimation results of the model.

4.2.2 Establishment of Health State Estimation Model Based on Health Indicators and Improved NSA-BP Model

Combining the establishment of ESP model, multi-cell model, electrochemical parameter identification algorithm, quantification methods of degradation modes,

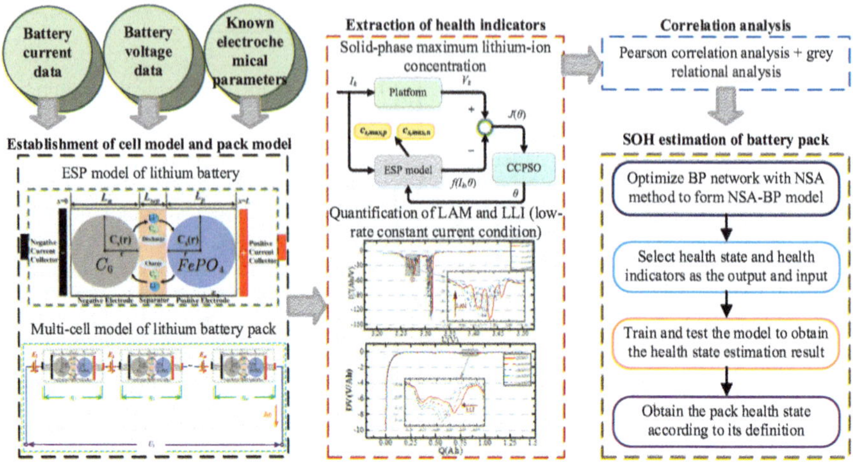

Fig. 4.7 The framework diagram of constructed health state estimation model for battery pack

correlation analysis methods, and neural network model, the framework diagram of health state estimation model for energy storage lithium-ion battery pack constructed in this book is shown in Fig. 4.7.

From Fig. 4.7, it can be seen that the framework of the energy storage lithium-ion battery pack health state estimation model proposed in this book is built based on the battery pack model, extraction of multidimensional health indicators, correlation analysis of health indicators, and the battery pack health state estimation network.

4.3 Summary

This chapter focuses on the study of health state estimation methods for energy storage lithium batteries. By studying the electrochemical parameters and degradation modes that are sensitive to the health state of the energy storage lithium battery, this topic forms a mapping relationship that takes multi-dimensional health indicators as the input of the NSA-BP neural network, and the health state of the energy storage lithium battery is the output of the NSA-BP neural network, so as to propose a new method for long-term health state estimation. This method employs a continuous adjustment of network parameters to minimize the discrepancy between the output health state estimation result and the expected value. Upon completion of model training, the health state value generated at that point is considered the final estimation result.

References

1. Xiong, R., et al. 2023. A novel nonlinear decreasing step-bacterial foraging optimization algorithm and simulated annealing-back propagation model for long-term battery state of health estimation. *Journal of Energy Storage* 59 (3): 106484–106495.
2. Yang, D., et al. 2017. A neural network based state-of-health estimation of lithium-ion battery in electric vehicles. *Energy Procedia* 105 (5): 2059–2064.
3. Dongdong, G., et al. 2022. Extreme learning machine using bat optimization algorithm for estimating state of health of lithium-ion batteries. *Applied Sciences* 12 (3): 1398–1407.
4. Shunli, W., et al. 2022. A critical review of improved deep convolutional neural network for multi-timescale state prediction of lithium-ion batteries. *Energies* 15 (14): 5053–5072.
5. Xin Sui, S.H., Søren B. Vilsen, Jinhao Meng, Remus Teodorescu, and Daniel-Ioan Stroe. 2021. A review of non-probabilistic machine learning-based state of health estimation techniques for lithium-ion battery. *Applied Energy* 300(15): 117346–117354.
6. Tian, H., et al. 2020. A review of the state of health for lithium-ion batteries: Research status and suggestions. *Journal of Cleaner Production* 261 (10): 120813–120821.
7. Chen, M., et al. 2023. An overview of data-driven battery health estimation technology for battery management system. *Neurocomputing* 1 (1): 1–9.
8. Kheirkhah-Rad, E., and M. Moeini-Aghtaie. 2021. A novel data-driven SOH prediction model for lithium-ion batteries. In *2021 31st Australasian universities power engineering conference (AUPEC)*, vol. 1, issue 1, 1–8.
9. Wen, J., et al. 2022. SOH prediction of lithium battery based on IC curve feature and BP neural network. *Energy* 261 (15): 125234–125244.

Chapter 5
Experimental Verification and Analysis of Health State Estimation for Lithium-Ion Battery Pack

Abstract The purpose of this chapter is to extract health indicators, validate ESP cell model and battery pack model, and validate health state estimation methods under specific experimental conditions. This chapter firstly establishes an experimental platform and clusters two major types of working conditions based on actual energy storage conditions. Secondly, this chapter extracts and analyzes the correlation of health indicators under typical energy storage conditions. Finally, this chapter validates the ESP model and multi-cell model under low-rate constant current energy storage degradation conditions and variable-rate energy storage degradation conditions, and conducted validation analysis of the health state estimation method on this basis. The analysis results demonstrate that the health state estimation method for energy storage lithium-ion battery packs developed in this book based on multidimensional health indicators and NSA-BP model has obvious advantages.

Keywords Energy storage degradation conditions · Validation analysis · Health state estimation · Health indicators · Pack models

5.1 Experimental Conditions and Experimental Platform

5.1.1 Clustering of Typical Energy Storage Conditions

Due to the fact that the simulated environment in this experiment is a typical energy storage environment, the experimental conditions for all cells and battery packs in this book should be typical energy storage conditions. According to the obtained data of the actual working conditions of a certain energy storage station, it can be seen that the working conditions of the energy storage station vary under different needs. The specific energy storage typical working conditions obtained are shown in Fig. 5.1.

In Fig. 5.1, C1–C6 represent working conditions 1–6, respectively. From Fig. 5.1, it can be seen that there are differences in the characteristics between working conditions, so typical energy storage working conditions should be clustered. The purpose of clustering is to merge working conditions with similar features by searching for

Q. Huang et al., *Long-Term Health State Estimation of Energy Storage Lithium-Ion Battery Packs*, https://doi.org/10.1007/978-981-99-5344-8_5

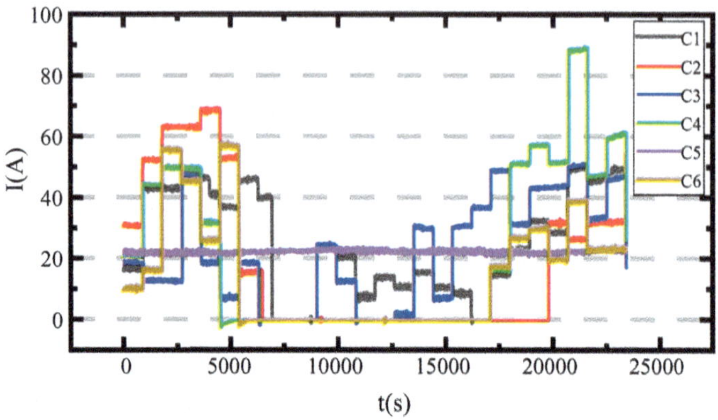

Fig. 5.1 Actual working conditions of an energy storage station

internal connections between samples. Before clustering, it is necessary to perform dimensionless processing on the data. This book chooses the same mean method as in the GRA method. Besides, this book selects the K-means method to cluster typical energy storage conditions. The optimization objective of the K-means algorithm is to minimize the sum of squared error (SSE) and maximize the silhouette coefficient (SC). The expression for the SSE is shown in Eq. (5.1).

$$\begin{cases} SSE = \sum_{j=1}^{K} \sum_{x \in C_j} \left\| x - \mu_j \right\|_2^2 \\ \mu_j = \frac{1}{|C_j|} \sum_{x \in C_j} x \end{cases} \tag{5.1}$$

In Eq. (5.1), x represents the sample coordinates, j represents the type of cluster, K represents the total number of cluster classes, C represents the cluster, and μ_j represents the cluster class center of cluster C_j. When applying the K-means method for clustering, the number of clusters needs to be given a predetermined value in advance. In addition, another indicator for evaluating the superiority of K-means method clustering results in this book is to maximize the SC. Smaller SSE and larger SC can reflect finer sample partitioning and better clustering results. The expression for the SC is shown in Eq. (5.2).

$$\begin{cases} S = \frac{1}{n} \sum_{i=1}^{n} S(i) \\ S(i) = \frac{b(i) - a(i)}{\max\{a(i), b(i)\}} \end{cases} \tag{5.2}$$

In Eq. (5.2), $a(i)$ represents the mean distance from sample i to other samples in the cluster, $b(i)$ represents the minimum mean distance from sample i to all samples in the neighboring cluster, $S(i)$ represents the SC of sample i, and S represents the overall SC.

For the actual energy storage station working conditions obtained in this book, the working condition features used for clustering include the longest constant current time during discharge and the number of large current steps. Among them, all working condition curves need to be smoothed before clustering analysis can be performed. By combining the number of working conditions and the two types of features extracted above, a sample set of 6 × 2 can be formed, then the K-means method is used for clustering typical energy storage conditions. The SSE and the SE which vary with the number of clusters in the K-means method are shown in Fig. 5.2.

From Fig. 5.2, it can be seen that as the number of clusters increases, SSE continuously decreases. When the number of cluster changes from 1 to 2, the decrease in the SSE significantly decreases. Obviously, when the number of clusters is 2, it reaches the elbow of the SSE curve. In addition, the change in the SC does not show a linear relationship with the clustering value, but it can be seen that the SC is the highest when the number of clusters is 2. Although the SSE in Fig. 5.2 is slightly lower when the number of cluster classes is 3 than when the number of cluster classes is 2, it can be seen from the changes in the SC that the number of cluster classes in 2 is the best. Therefore, for typical energy storage conditions, the optimal number of clusters is 2. The specific clustering situation is shown in Table 5.1.

Through analysis, it can be concluded that the typical energy storage conditions after clustering can be divided into low-rate constant current energy storage conditions and variable-rate energy storage conditions. Due to the need for cyclic degradation experiments on the individual or cluster of energy storage lithium-ion batteries in this book, all experimental conditions of energy storage lithium-ion batteries in

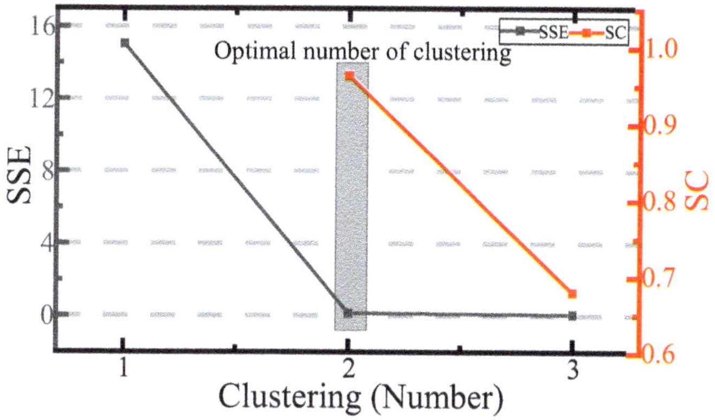

Fig. 5.2 SSE and SE of typical energy storage conditions under different cluster numbers

Table 5.1 Clustering results of typical energy storage conditions

Number of clusters	Types of conditions
Cluster 1	C1–C4, C6
Cluster 2	C5

this book require cycling of the two clustered conditions. Specifically, in this experiment, smooth and proportionally reduced C5 and partial C6 are used as a low-rate constant current energy storage condition and a variable-rate energy storage condition, respectively.

5.1.2 Experimental Platform and Experimental Setup

The cells studied in this book are some new IFR26650 lithium iron phosphate battery with rated capacity of 2.3 Ah, upper limit cut-off voltage of 3.6 V, and lower limit cut-off voltage of 2.5 V. In order to verify the superiority of the estimation algorithm proposed in this book, all operational steps are strictly executed under a battery high-rate charge-discharge tester (BTS750-200-100-4) and a temperature control thermostat (BTKS5-150C). In addition, the data obtained from the experiment is transmitted to the host through a charge discharge tester for storage and use. The specific structure of the experimental platform is shown in Fig. 5.3.

In this book, the temperature of the thermostat under all experimental conditions is set to 25 °C. The host configurations for analyzing and processing data are Intel Core i5-9500 and NVIDIA GeForce GT 710. The experiments to be performed include the 0.1 C constant current full discharge test of *Cell*1, the 0.4 C constant current energy storage degradation test of *Cell*2, the 0.5 C constant current energy storage degradation test of *Pack*1, and the variable-rate energy storage degradation test of *Pack*2. Among them, each battery pack is obtained by connecting four individual cells in series. C represents the rate of battery charging and discharging capacity.

The data of Cell1 is used to obtain the expression of positive open-circuit voltage, the data of Cell2 is used to train the health state estimation model, and the data of Pack1 and Pack2 are used to test and analyze the health state estimation model. The current curves within one cycle in the constant current energy storage degradation

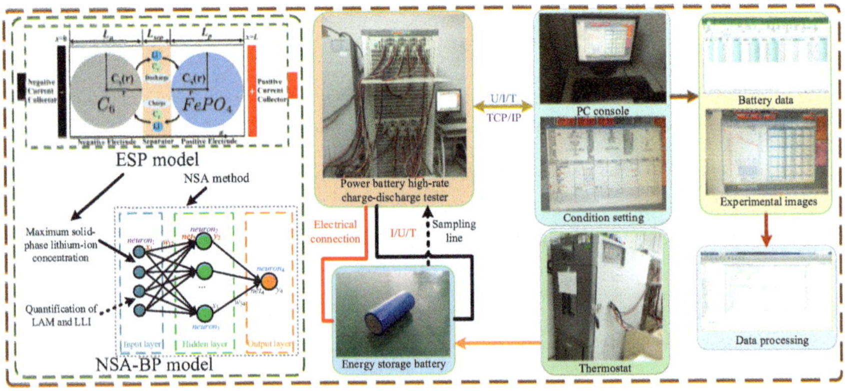

Fig. 5.3 The specific structure of the experimental platform

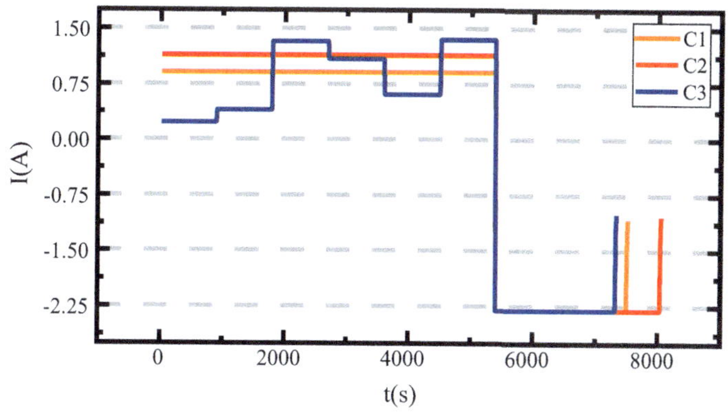

Fig. 5.4 The current curves within one degradation cycle

experiments and the variable-rate energy storage degradation experiment is shown in Fig. 5.4.

In Fig. 5.4, C1 represents the 0.4 C constant current energy storage degradation condition, C2 represents the 0.5 C constant current energy storage degradation condition, and C3 represents the variable-rate energy storage degradation condition. All three conditions are during constant current and constant voltage charging. The current is negative during charging and positive during discharge. Unlike the open-source battery degradation datasets, the energy storage lithium-ion battery in this experiment requires obtaining the maximum dischargeable capacity of the battery every time it undergoes a degradation cycle. In open-source datasets, the degradation conditions of batteries are all low-rate constant current full discharge conditions. In this kind of working conditions, the maximum discharge capacity of the battery under each degradation cycle can be accurately calculated through the collected current and time data. However, the degradation conditions of *Cell*2, *Pack*1 and *Pack*2 in this book do not belong to the low-rate constant current full discharge cycle condition, and the maximum discharge capacity of each cycle cannot be obtained through data collection. Therefore, in order to obtain the maximum discharge capacity of the battery, it is necessary to conduct capacity measurement experiments on each cell. In this book, *Cell*2, *Pack*1 and *Pack*2 undergo a corresponding capacity measurement experiment every five days of degradation.

5.2 Health Indicators Extraction Results and Correlation Analysis Results

5.2.1 Identification Results of Electrochemical Parameters

In order to establish a specific ESP model to identify parameters, it is necessary to obtain an expression for the open-circuit voltage of the positive and negative electrodes. The expression for the open-circuit voltage of the negative electrode is empirical [1]. Through low-rate constant current experiments, the expression of the positive electrode open-circuit voltage can be obtained by combining empirical expressions and the difference between the simulated terminal voltage of the SP model [2]. Therefore, the expressions of positive and negative open-circuit voltage of energy storage lithium-ion battery are shown in Eq. (5.3).

$$\begin{cases} E_n(\theta_n) = 0.6379 + 0.5416e^{-305.5309\theta_n} + 0.044\tanh\left(\frac{-(\theta_n-0.1958)}{0.1088}\right) - 0.1978\tanh\left(\frac{\theta_n-1.0571}{0.0854}\right) \\ -0.6875\tanh\left(\frac{\theta_n+0.0117}{0.0529}\right) - 0.0175\tanh\left(\frac{\theta_n-0.5692}{0.0875}\right) \\ E_p(\theta_p) = 509.0485\theta_p^8 - 2584.6784\theta_p^7 + 5232.1118\theta_p^6 - 5526.4540\theta_p^5 + 3311.6219\theta_p^4 \\ -1138.2716\theta_p^3 + 214.7627\theta_p^2 - 19.7519\theta_p + 4.0857 \end{cases} \tag{5.3}$$

According to Eq. (5.3), the curves of the open-circuit voltage of the positive and negative electrodes of the energy storage lithium-ion battery varying with the utilization rate of the positive and negative electrodes are shown in Fig. 5.5.

In Fig. 5.5, a, b show the open-circuit voltage curves of the positive and negative electrodes as the functions of positive and negative electrode utilization, respectively. In addition to the positive and negative open-circuit voltage expressions, the ESP model also needs to obtain some constants, as well as some internal electrochemical parameters and structural parameters of the battery. Some relevant parameter values in the energy storage lithium-ion battery ESP model used in this book are shown in Table 5.2.

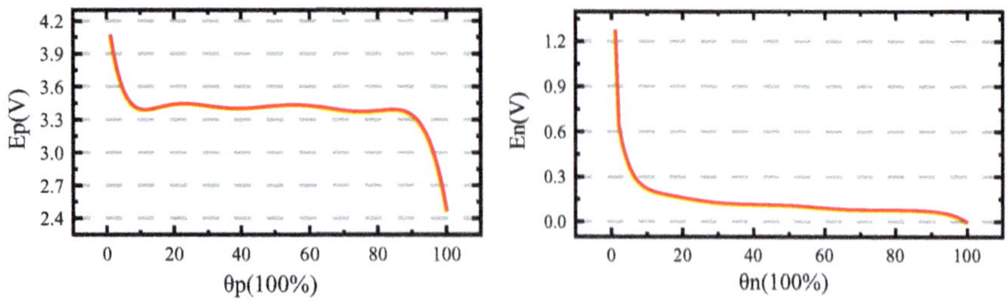

(a) The open-circuit voltage of the positive electrode

(b) The open-circuit voltage of the negative electrode

Fig. 5.5 The open-circuit voltage of the positive and negative electrodes

Table 5.2 Some relevant parameter values in the ESP model

Parameter type	Symbol	Value (25 °C)
Faraday constant (C/mol)	F	96,487
Universal gas constant (J/mol/K)	R	8.314
Liquid-phase lithium-ion migration coefficient	t_+	0.363
Battery temperature (K)	T	298.15
Liquid-phase lithium-ion conductivity	$\kappa_p/\kappa_n/\kappa_{sep}$	0.265/0.183/0.168
Plate thickness (m)	$L_p/L_n/L_{sep}$	$70 \times 10^{-6}/34 \times 10^{-6}/16 \times 10^{-6}$
Effective area of pole piece (m^2)	A_p/A_n	0.17/0.17
The radius of the active particle (m)	$R_{s,p}/R_{s,n}$	$3.65 \times 10^{-8}/3.5 \times 10^{-6}$
Solid-phase diffusion coefficient (m^2/s)	$D_{s,p}/D_{s,n}$	$1.18 \times 10^{-18}/2 \times 10^{-14}$
Liquid-phase diffusion coefficient (m^2/s)	$D_{e,p}/D_{e,n}$	$4.97 \times 10^{-9}/4.97 \times 10^{-9}$
SEI film Ohmic resistance (Ω m^2)	$R_{SEI,p}/R_{SEI,n}$	0.001/0.001
Solid-phase volume fraction	$\varepsilon_{s,p}/\varepsilon_{s,n}$	0.56/0.5
Liquid-phase volume fraction	$\varepsilon_{e,p}/\varepsilon_{e,n}/\varepsilon_{e,sep}$	0.3/0.3/1
Average electrode reaction rate constant ($m^{2.5}mol^{-0.5}s^{-1}$)	k_p/k_n	$3 \times 10^{-11}/8.19 \times 10^{-12}$
Initial liquid-phase lithium-ion concentration (mol/m^3)	$c_{e,0,p}/c_{e,0,n}$	1000/1000
Initial surface solid-phase lithium-ion concentration (mol/m^3)	$c_{s,surf,0,p}/c_{s,surf,0,n}$	3900/14870

The expression of the battery terminal voltage that only concerns the maximum solid-phase lithium-ion concentrations of the positive and negative electrodes can be obtained by substituting the actual or reference values of the parameters shown in Table 5.2 into the ESP cell model of the energy storage lithium-ion battery. In this book, the standard PSO algorithm and the CCPSO algorithm are used to obtain and compare the identification results of the maximum solid-phase lithium-ion concentrations of the positive and negative electrodes, and the results can be used to illustrate the superiority of the improved CCPSO algorithm. In this book, the assignment of relevant parameters in the standard PSO algorithm and the CCPSO algorithm is shown in Table 5.3.

Since *Cell*1 only conducts a 0.1 C constant current full charge and discharge test experiment, and the main purpose of this experiment on *Cell*1 is to help obtain the expression of positive open-circuit voltage of energy storage lithium-ion battery, it is not necessary to conduct electrochemical parameter identification, quantification of degradation modes, correlation analysis of health indicators, battery model verification and health state estimation method verification on *Cell*1. The electrochemical parameters of *Cell*2, *Pack*1, and *Pack*2 are used for subsequent training and testing

Table 5.3 The relevant parameters in the standard PSO algorithm and the CCPSO algorithm

Parameter type	Symbol	Value
The maximum solid-phase Li-ion concentration of positive electrode (mol/m^3)	$c_{s,\max,p}$	[16000,27000]
The maximum solid-phase Li-ion concentration of negative electrode (mol/m^3)	$c_{s,\max,n}$	[25000,36000]
Parameter dimension	d	2
Maximum iterations	k_{iter}	100
Initial speed of particles in each dimension	$v_{i,0}$	1000 ± 100
The maximum value of weight factor in standard PSO algorithm	$w_{\max}/w_{\min}$	0.9/0.4
The maximum weight factor of the first sub-particle in CCPSO algorithm	$w_{\max,1}/w_{\min,1}$	0.9/0.7
The maximum weight factor of the second sub-particle in CCPSO algorithm	$w_{\max,2}/w_{\min,2}$	0.6/0.4
Learning factor	c_1/c_2	2/2
Number of subpopulation categories	M	4
Particle swarm size of standard PSO algorithm	N_1	20
Particle swarm size of CCPSO algorithm	N_2	10

of neural network model, so the relevant identification process should be described in detail.

In this book, the standard PSO algorithm is firstly used to identify the electrochemical parameters of the ESP model. However, the convergence value of the identification target fitness is sometimes large, and the results of multiple parameter identification under the same degradation cycle are quite different. Therefore, in order to reduce the computational errors caused by the above shortcomings, this book develops a CCPSO algorithm for improvement. In order to highlight the improvement effect of the CCPSO algorithm, this book compares and analyzes the change of the fitness value of the identification target of the algorithm and the standard PSO algorithm. Since there are several cells in each pack, one cell in *Pack*1 and one cell in *Pack*2 are randomly selected for fitness value analysis, respectively. In *Cell*2 and two cells randomly selected from *Pack*1 and *Pack*2, the identification target fitness values of the standard PSO algorithm and the CCPSO algorithm are used for the first degradation cycle and the last degradation cycle, as shown in Fig. 5.6.

In Fig. 5.6, S1 and S2 represent the CCPSO algorithm and the standard PSO algorithm, respectively. It can be seen that in the process of convergence to the optimal solution, the performance results of the CCPSO algorithm are better than that of the standard PSO algorithm, and the optimal fitness values of the CCPSO algorithm are obviously smaller. In Fig. 5.6c, the fitness of the algorithm converges after 54 iterations and 38 iterations, respectively, and the corresponding fitness values are 0.025301 and 0.020652, respectively, which shows that the improved algorithm is superior. The above discussion verifies the superiority of the CCPSO algorithm.

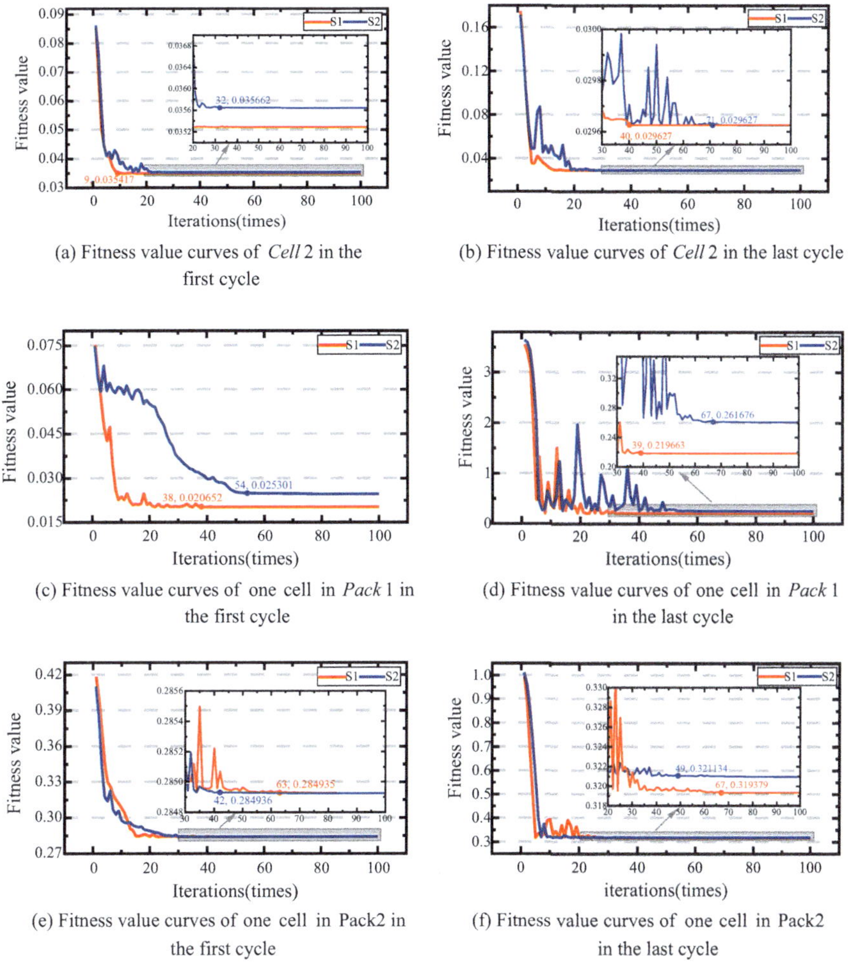

(a) Fitness value curves of *Cell* 2 in the first cycle

(b) Fitness value curves of *Cell* 2 in the last cycle

(c) Fitness value curves of one cell in *Pack* 1 in the first cycle

(d) Fitness value curves of one cell in *Pack* 1 in the last cycle

(e) Fitness value curves of one cell in Pack2 in the first cycle

(f) Fitness value curves of one cell in Pack2 in the last cycle

Fig. 5.6 Comparison curves of fitness value

During the continuous degradation of health state in the energy storage lithium-ion battery pack to 80% capacity, the electrochemical parameter identification results of *Cell*2, each cell in *Pack*1 and each cell in *Pack*2 obtained using the CCPSO algorithm are shown in Fig. 5.7.

From Fig. 5.7, it can be seen that as the battery degrades, the electrochemical parameter identification results of each experimental cell both show a decreasing trend. In addition, it can be seen that there is a small inconsistency between the individual cells of the energy storage lithium-ion battery in the same pack, resulting in a slight difference in the degradation between the cells. Among them, the difference in the degradation stage of energy storage lithium-ion batteries is significantly greater than that of new batteries.

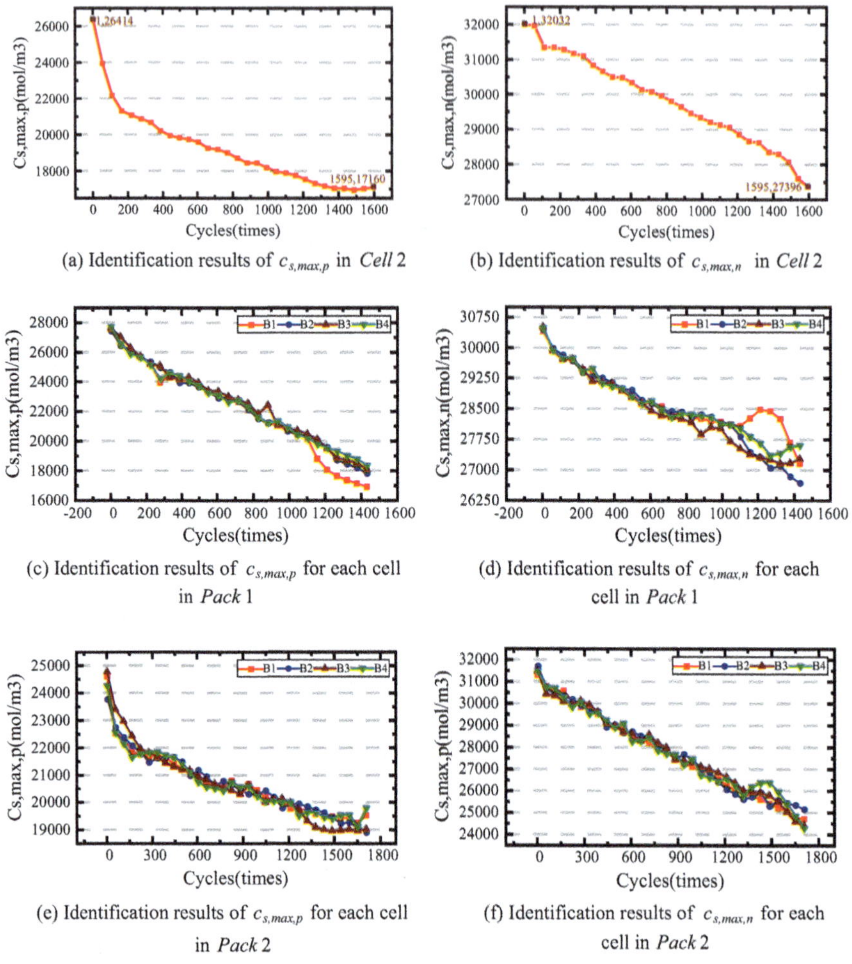

(a) Identification results of $c_{s,max,p}$ in *Cell* 2

(b) Identification results of $c_{s,max,n}$ in *Cell* 2

(c) Identification results of $c_{s,max,p}$ for each cell in *Pack* 1

(d) Identification results of $c_{s,max,n}$ for each cell in *Pack* 1

(e) Identification results of $c_{s,max,p}$ for each cell in *Pack* 2

(f) Identification results of $c_{s,max,n}$ for each cell in *Pack* 2

Fig. 5.7 The electrochemical parameter identification results

5.2.2 Quantitative Results of Degradation Modes

The features on IC curves and DV curves are used to quantify LAM and LLI, respectively. In this book, constant ΔU in IC curves is taken as 0.001 V, and constant ΔQ in DV curves is taken as 0.002 Ah. During the degradation of *Cell*2 and *Pack*1, the LAM quantification result curves and LLI quantification result curves of each cell are shown in Fig. 5.8.

The quantification results of LAM and LLI are shown in Fig. 5.8. LAM and LLI are both the main causes of battery degradation. From Fig. 5.8, it can be seen that during the degradation process of the battery, the quantification results of LAM and LLI are not significantly different in numerical terms, and both show an increasing

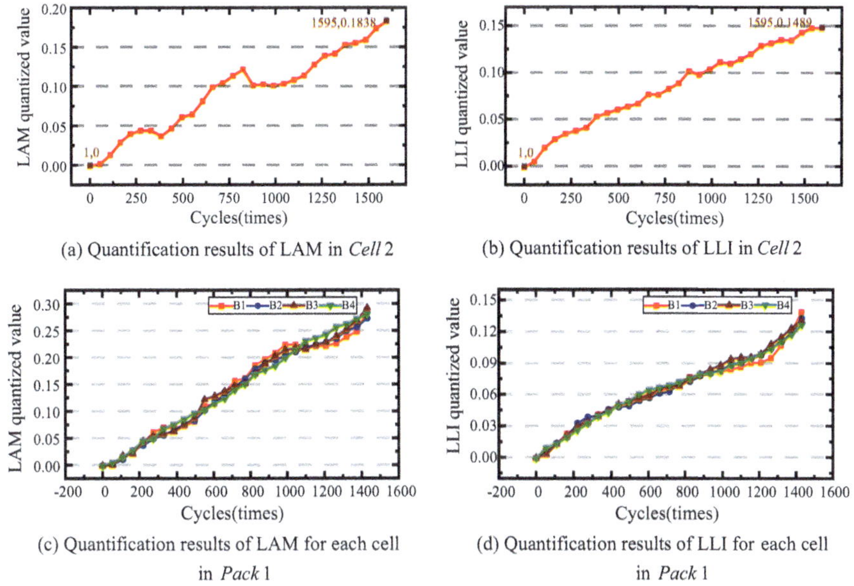

(a) Quantification results of LAM in *Cell* 2

(b) Quantification results of LLI in *Cell* 2

(c) Quantification results of LAM for each cell in *Pack* 1

(d) Quantification results of LLI for each cell in *Pack* 1

Fig. 5.8 The LAM quantification result curves and LLI quantification result curves

trend in the later stage of battery degradation. Due to the opposite trend between the quantification curves of LAM and LLI and the capacity degradation trend, negative values are taken for correlation analysis when calculating correlation and estimating health state using neural network models.

5.2.3 Correlation Analysis Results of Health Indicators

The initial maximum capacity of *Cell*2 is 2.287 Ah, the initial maximum capacities of each cell in *Pack*1 are 2.293, 2.289, 2.297, and 2.306 Ah, respectively, and the initial maximum capacities of each cell in *Pack*2 are 2.294, 2.305, 2.291, and 2.288 Ah, respectively. After experiencing 1595, 1430, and 1705 degradation cycles, respectively, the battery capacity of *Cell*2, *Pack*1, and *Pack*2 degrades to just under 80% of the rated capacity. When the battery capacity degrades to below 80% of the rated capacity, the degradation experiment of the cell or pack is stopped. During the degradation, the capacity degradation curve of each experimental cell can be obtained through multiple capacity testing experiments, and the health state degradation curve of each experimental cell can be obtained using the capacity definition method of cell health state. The capacity degradation curves and health state degradation curves of each cell in *Cell*2, *Pack*1, and *Pack*2 are shown in Fig. 5.9.

In order to analyze the correlation or relation between health indicators and battery health state curves, this book uses the curves of the above four health indicators as

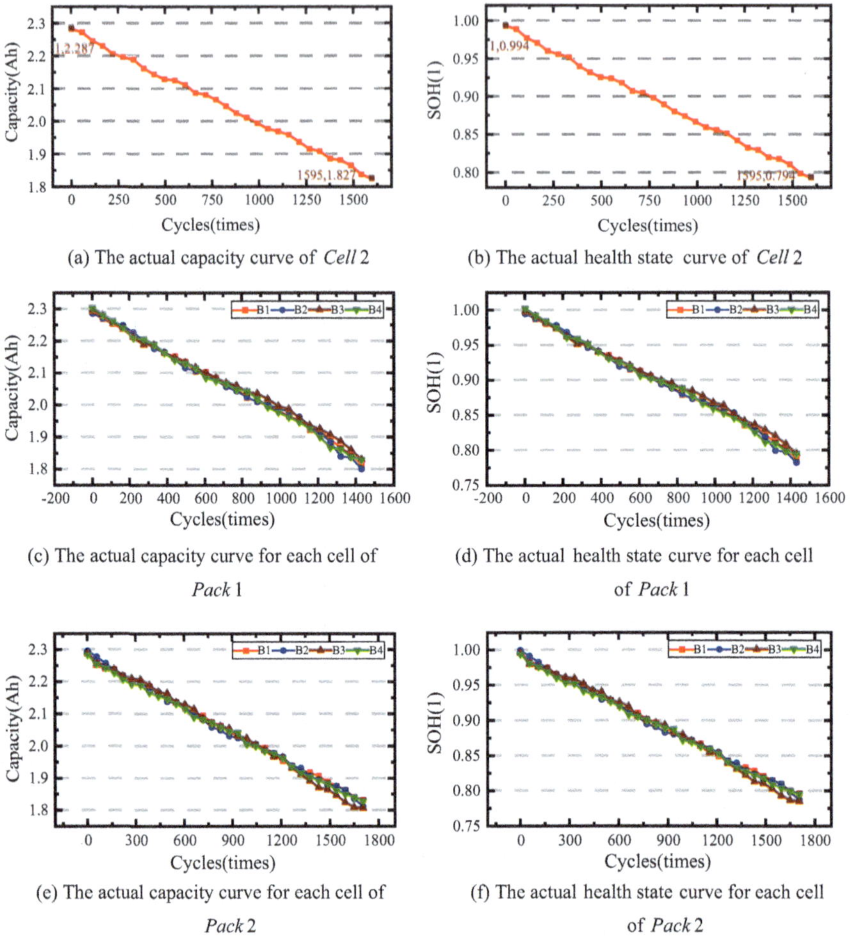

(a) The actual capacity curve of *Cell* 2

(b) The actual health state curve of *Cell* 2

(c) The actual capacity curve for each cell of *Pack* 1

(d) The actual health state curve for each cell of *Pack* 1

(e) The actual capacity curve for each cell of *Pack* 2

(f) The actual health state curve for each cell of *Pack* 2

Fig. 5.9 The capacity degradation curves and health state degradation curves

comparative sequences, and the battery health state curve as the reference sequence. PCA and GRA methods are used to analyze the correlation coefficients and relational degrees between the sequences to evaluate the correlation between each health indicator and battery capacity. In the GRA method, the resolution coefficient is taken as 0.5. The Pearson correlation coefficients and grey relational degrees between the above health indicators and the corresponding cell health state in *Cell*2, *Pack*1, and *Pack*2 are shown in Table 5.4.

In Table 5.4, the data in the first four rows and the last four rows correspond to the Pearson correlation coefficients and grey correlation degrees, respectively. It can be seen that the above health indicators are highly correlated with health state. It is worth noting that the above four health indicators can enhance the universality of the health state neural network estimation model. When voltage, current, and number

Table 5.4 The Pearson correlation coefficients and grey relational degrees

Para	*Cell*2	*Pack*1-1	*Pack*1-2	*Pack*1-3	*Pack*1-4	*Pack*2-1	*Pack*2-2	*Pack*2-3	*Pack*2-4
$c_{s,\max,p}$	0.9160	0.9900	0.9950	0.9939	0.9921	0.9329	0.9733	0.9361	0.9329
$c_{s,\max,n}$	0.9960	0.9219	0.9882	0.9807	0.9758	0.9990	0.9880	0.9921	0.9821
LLI	0.9955	0.9834	0.9915	0.9950	0.9914	/	/	/	/
LAM	0.9737	0.9836	0.9862	0.9901	0.9964	/	/	/	/
$c_{s,max,p}$	0.5016	0.7752	0.5999	0.7383	0.6805	0.5319	0.5866	0.5158	0.5529
$c_{s,max,n}$	0.8490	0.5534	0.6216	0.4661	0.5263	0.8756	0.6338	0.8131	0.8282
LLI	0.7665	0.7112	0.7048	0.7896	0.7198	/	/	/	/
LAM	0.8208	0.7708	0.6788	0.7916	0.8238	/	/	/	/

of cycles are used as health indicators, they can only meet the estimation of health state under a certain working condition. However, the models based on $c_{s,\max,p}$ and $c_{s,\max,n}$ can provide accurate estimates for various complex working conditions, and the models based on LLI and LAM can effectively estimate health state under various low-rate constant current working conditions.

5.3 Verification of Battery Model and Estimation Method Under Low-Rate Constant Current Energy Storage Degradation Conditions

5.3.1 Verification of ESP Model and Pack Model Under Low-Rate Constant Current Energy Storage Degradation Conditions

Firstly, the reference values of electrochemical parameters are used to model the cell under low-rate constant current energy storage degradation conditions. Due to the low-rate constant current energy storage degradation conditions of *Cell*2 and *Pack*1, *Cell*2 and one cell in *Pack*1 is selected for comparison of the simulation performance of the SP model and ESP model under those conditions. Under the first degradation cycle of *Cell*2 and one cell in *Pack*1, the actual terminal voltage and the simulated terminal voltage of two simulation cell models are shown in Fig. 5.10.

From Fig. 5.10, it is evident that both two models have good simulation results under low-rate conditions. Specifically, in Fig. 5.10, the MAEs of two terminal voltages are shown in Table 5.5.

In Table 5.5, the performance of the ESP model is better under low-rate conditions. During partial degradation cycles, the terminal voltage curves of *Cell*2, one cell in *Pack*1 and *Pack*1 are shown in Fig. 5.11.

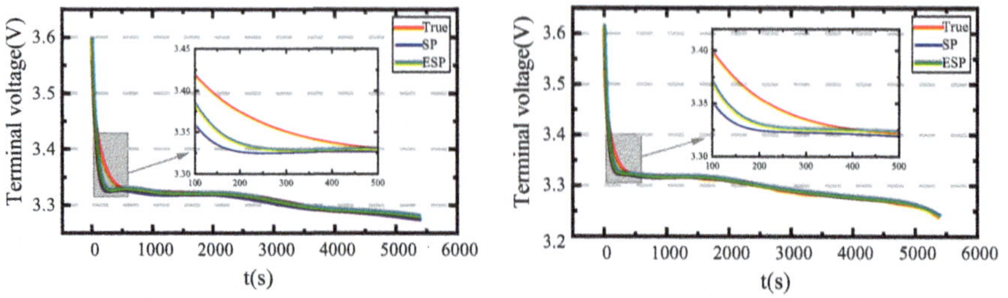

(a) The terminal voltages of *Cell* 2 in the first degradation cycle

(b) The terminal voltages of one cell in *Pack* 1 in the first degradation cycle

Fig. 5.10 The actual terminal voltage and the simulated terminal voltage

Table 5.5 The MAEs of these two model terminal voltages

Cell	*Cell*2 (SP model)	*Cell*2 (ESP model)	One cell in *Pack*1 (SP model)	One cell in *Pack*1 (ESP model)
MAE	0.0064 V	0.0036 V	0.0032 V	0.0030 V

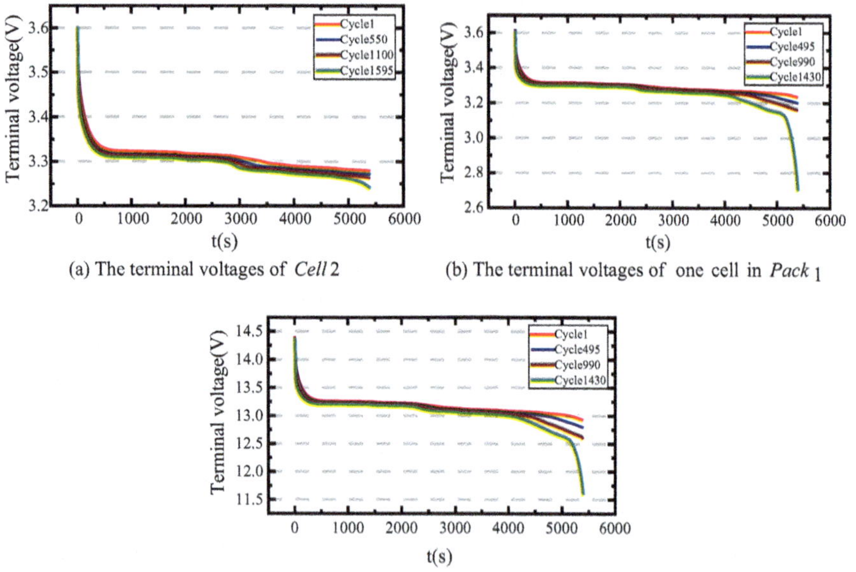

(a) The terminal voltages of *Cell* 2

(b) The terminal voltages of one cell in *Pack* 1

(c) The terminal voltages of *Pack* 1

Fig. 5.11 The measured terminal voltages of *Cell*2 and *Pack*1

Due to the large number of degradation cycles, this book only selects the first and last cycles of degradation for model validation. The terminal voltages and their errors of the relevant cells or packs are shown in Fig. 5.12. It is worth noting that in Figs. 5.10 and 5.12, the maximum solid-phase lithium-ion concentrations in the

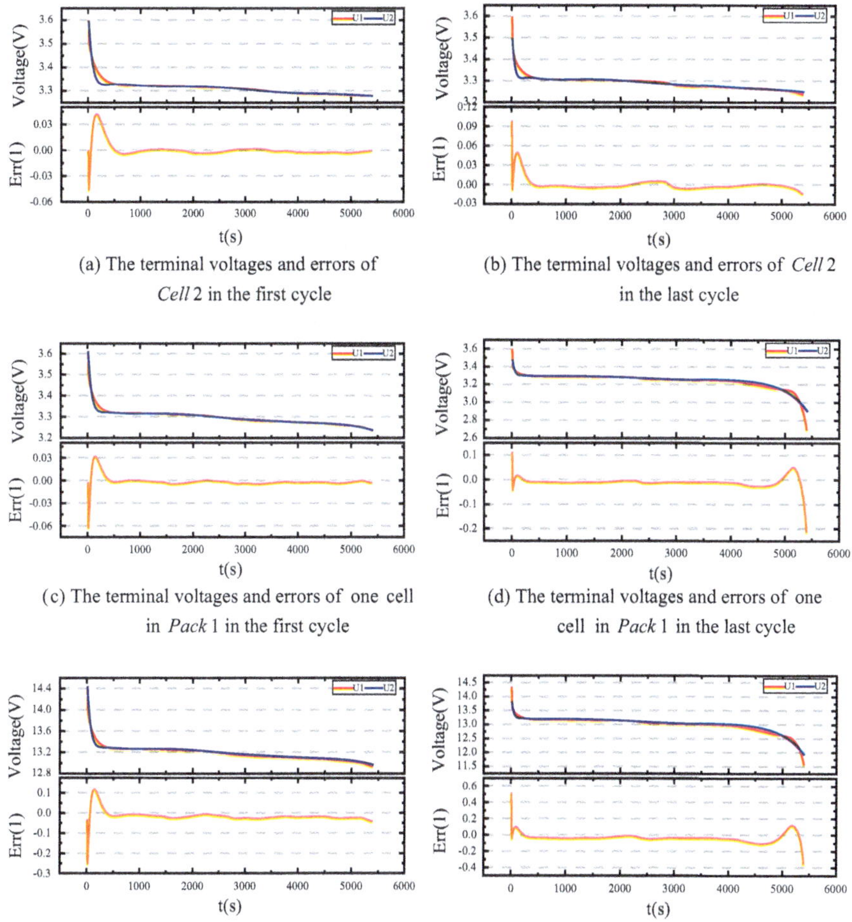

(a) The terminal voltages and errors of *Cell* 2 in the first cycle

(b) The terminal voltages and errors of *Cell* 2 in the last cycle

(c) The terminal voltages and errors of one cell in *Pack* 1 in the first cycle

(d) The terminal voltages and errors of one cell in *Pack* 1 in the last cycle

(e) The terminal voltages and errors of *Pack* 1 in the first cycle

(f) The terminal voltages and errors of *Pack* 1 in the last cycle

Fig. 5.12 Model verification curves of *Cell*2 and *Pack*1

positive and negative electrodes of the simulation model are the identification results of the CCPSO algorithm and reference values, respectively.

Figure 5.12 displays the actual terminal voltage ($U1$) and simulated terminal voltage ($U2$) of ESP model or ESP-MCM. In Fig. 5.12, a, b or c, d or e, f offer a comparison between the new cell and pack and the degraded cell and pack, demonstrating a significant decrease in voltage after the same time of discharge. This decrease suggests a reduction in capacity performance due to battery degradation. Table 5.6 shows the mean absolute errors (MAEs) of terminal voltages in *Cell*2, one cell of *Pack*1 and *Pack*1.

Figure 5.12 and Table 5.6 demonstrate that ESP model and ESP-MCM exhibit minimal errors under two low-rate constant current energy storage degradation

Table 5.6 The MAEs of terminal voltages in *Cell*2, one cell of *Pack*1 and *Pack*1

Cell/pack (voltage model)	MAE of the first cycle (V)	MAE of the last cycle (V)
*Cell*2 (ESP model)	0.0034	0.0043
One cell in *Pack*1 (ESP model)	0.0026	0.0110
*Pack*1 (ESP-MCM)	0.0187	0.0366

conditions. Specifically, the MAEs of terminal voltage in the first and last cycles are compared in Table 5.6, revealing that the accuracy of ESP model and ESP-MCM slightly decreases with battery degradation, but remains within the acceptable range. These findings confirm the effectiveness and superiority of the developed ESP model and ESP-MCM under low-rate constant current energy storage degradation conditions.

5.3.2 Health State Estimation Method Validation Under Low-Rate Constant Current Energy Storage Degradation Condition

In this study, the researchers analyze the results of health state estimation under two low-rate constant current energy storage degradation conditions and one variable-rate energy storage degradation condition. Specifically, $c_{s,\max,p}$, $c_{s,\max,n}$, and the quantification results of LAM and LLI are regarded as inputs for a BP neural network, with the health state sequence as the output. To evaluate the impact of the quantification results on the neural network, another BP neural network is trained and tested using only $c_{s,\max,p}$ and $c_{s,\max,n}$ as inputs. Additionally, they trained and tested the health state estimation model using the four HIs proposed in this study, with the BP neural network, SA-BP neural network, and NSA-BP neural network employed for this purpose. Table 5.7 presents the initial parameter settings for these neural networks.

In this book, all HI sequences and health state degradation sequences of *Cell*2 are used as the training set for the neural network model. When using data from *Cell*2 for training, if the number of iterations exceeds the set value or the target mean square error meets the requirements, the training of the model is stopped. The training process of different neural network models is shown in Fig. 5.13.

In Fig. 5.13, E1, E2, E3, and E4 represent BP neural networks based on electrochemical parameters, BP neural networks based on electrochemical parameters and degradation modes, SA-BP neural networks based on electrochemical parameters and degradation modes, and NSA-BP neural networks based on electrochemical parameters and degradation modes, respectively. From Fig. 5.13, it can be seen that different neural network models need to train each individual number of times to meet the set target MAE requirements. The network connection weights and thresholds are recorded and used as model parameters for the test set. In this section, the

Table 5.7 The initial parameter settings for neural networks

Parameter	Symbol	Value/type
Initial/final temperature (°C)	T_0/T_f	8/3
Temperature decay scale of SA	λ	0.85
Maximum iterations of NSA	K_{max}	42
Step factor	c	0.2
Markov length	L	10
Number of input/hidden/output layer nodes	/	2/5/1
Learning rule	/	L-M
Learning rate	/	0.01
Target mean squared error	/	1×10^{-3}
The activation function of the hidden/output layer	/	Sigmoid/purelin

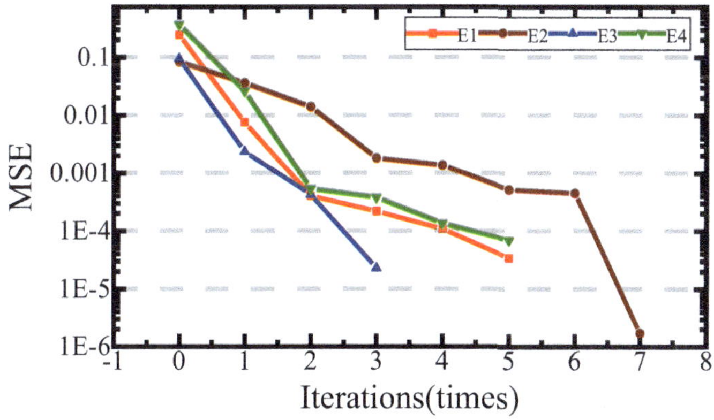

Fig. 5.13 The training process of four different neural network models

relevant data of *Pack*1 is regarded as the test set. The diagrams of health state output from the test set and its error are shown in Fig. 5.14.

Figure 5.14 depicts the actual health state ($S1$), the health state based on BP network and $c_{s,\max,i}$ ($S2$), the health state based on BP network and four HIs ($S3$), the health state based on simulated annealing-back propagation (SA-BP) network and four HIs ($S4$), and the health state based on NSA-BP network and four HIs ($S5$). $Err1$, $Err2$, $Err3$, and $Err4$ represent the same meaning corresponding to $S2$, $S3$, $S4$ and $S5$, respectively. The performance evaluation indicators for these algorithms in *Pack*1 are presented in Table 5.8.

From Table 5.8, it can be seen that the health state estimation algorithm based on multidimensional HIs and NSA-BP model is obviously better than the previous algorithm in terms of RMSE and MAE error indicators, both for each cell in *Pack*1 and for the entire *Pack*1.

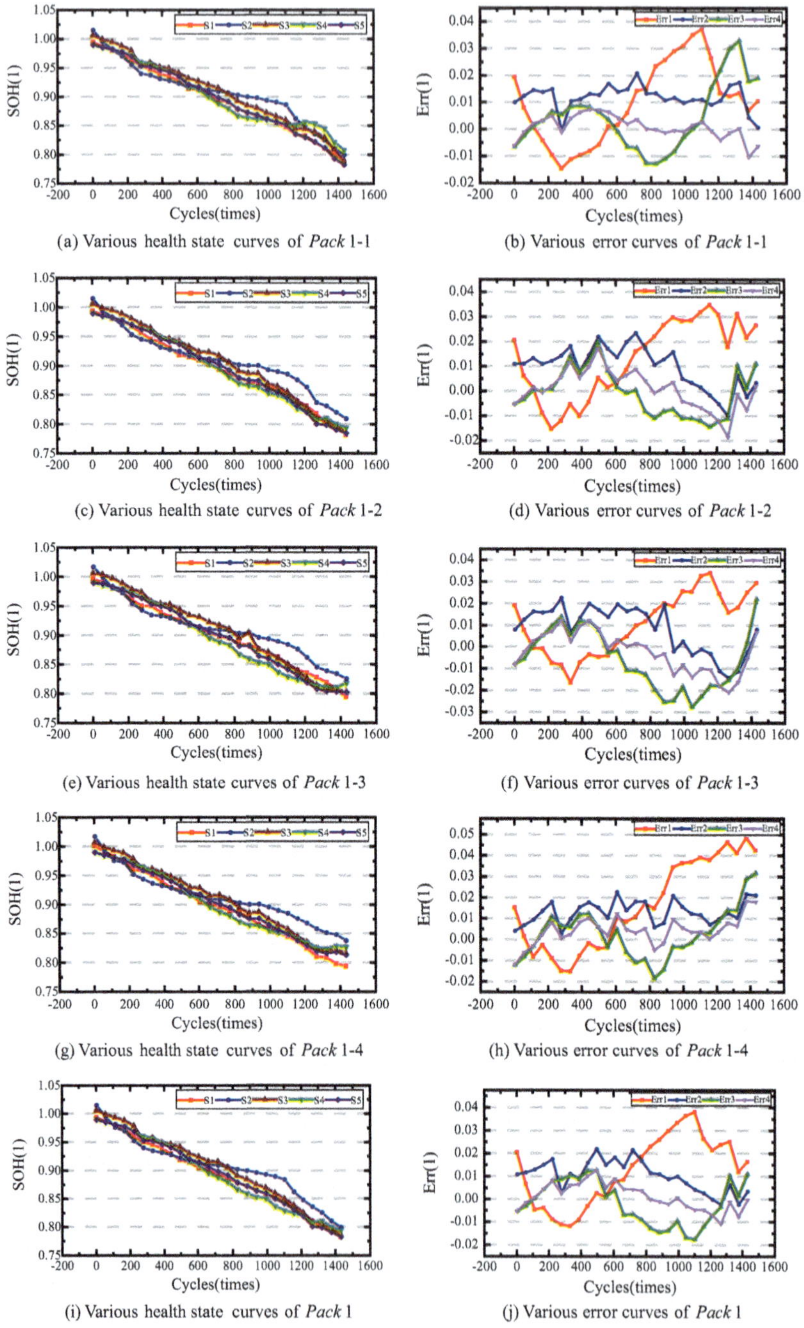

(a) Various health state curves of *Pack* 1-1

(b) Various error curves of *Pack* 1-1

(c) Various health state curves of *Pack* 1-2

(d) Various error curves of *Pack* 1-2

(e) Various health state curves of *Pack* 1-3

(f) Various error curves of *Pack* 1-3

(g) Various health state curves of *Pack* 1-4

(h) Various error curves of *Pack* 1-4

(i) Various health state curves of *Pack* 1

(j) Various error curves of *Pack* 1

Fig. 5.14 The diagrams of health state output from *Pack*1 and its error

Table 5.8 The performance evaluation indicators for algorithms in *Pack*1

Type	*Pack*1-1	*Pack*1-2	*Pack*1-3	*Pack*1-4	*Pack*1
*S*2	RMSE: 0.0181 MAE: 0.0149	RMSE: 0.0207 MAE: 0.0176	RMSE: 0.0181 MAE: 0.0148	RMSE: 0.0268 MAE: 0.0215	RMSE: 0.0199 MAE: 0.0167
*S*3	RMSE: 0.0130 MAE: 0.0122	RMSE: 0.0131 MAE: 0.0115	RMSE: 0.0137 MAE: 0.0120	RMSE: 0.0146 MAE: 0.0135	RMSE: 0.0125 MAE: 0.0108
*S*4	RMSE: 0.0129 MAE: 0.0098	RMSE: 0.0093 MAE: 0.0078	RMSE: 0.0147 MAE: 0.0126	RMSE: 0.0122 MAE: 0.0097	RMSE: 0.0097 MAE: 0.0086
*S*5	RMSE: 0.0044 MAE: 0.0034	RMSE: 0.0076 MAE: 0.0062	RMSE: 0.0094 MAE: 0.0077	RMSE: 0.0080 MAE: 0.0066	RMSE: 0.0059 MAE: 0.0048

5.4 Verification of Battery Model and Estimation Method Under Variable-Rate Constant Current Energy Storage Degradation Condition

5.4.1 Verification of ESP Model and Pack Model Under Variable-Rate Constant Current Energy Storage Degradation Condition

This section firstly uses electrochemical parameter reference values, including the maximum solid-phase lithium-ion concentrations in the positive and negative electrodes, to model SP model under variable-rate energy storage degradation conditions, but the simulation effect is poor. To enhance model simulation capability, this book proposes an ESP model, and on this basis, establishes a multi-cell pack model. The curves of the actual terminal voltage and two simulated terminal voltages during the first degradation cycle of one cell in *Pack*2 are shown in Fig. 5.15.

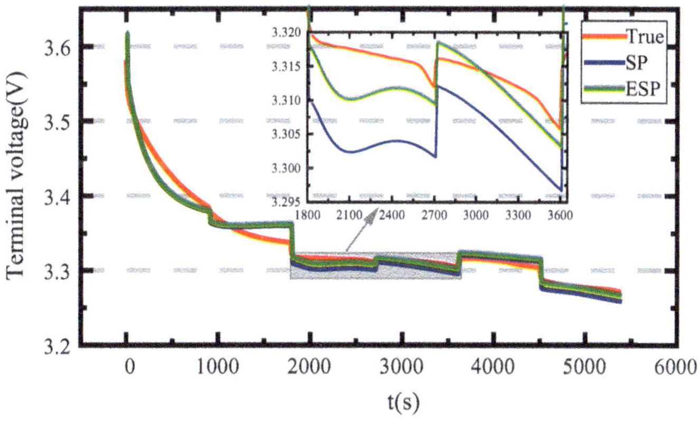

Fig. 5.15 The curves of terminal voltages during the first cycle of one cell in *Pack*2

From Fig. 5.15, it can be seen that under the variable-rate energy storage degradation condition, the simulation effect of the ESP model is better than that of the SP model. The ESP model and SP model terminal voltage MAE of one cell in *Pack*2 are shown in Table 5.9.

In the experiment in this book, *Pack*2 is in a variable-rate energy storage degradation condition. One cell in *Pack*2 and *Pack*2 are used for ESP model validation and ESP-MCM multi-cell model validation under the variable-rate energy storage degradation condition, respectively. The terminal voltage curves of one cell in *Pack*2 and entire *Pack*2 under partial degradation cycles are shown in Fig. 5.16.

Due to the large number of degradation cycles, this book only selects the first and last cycles of degradation for model validation. The terminal voltages and their errors of the relevant cells or packs are shown in Fig. 5.17. It is worth noting that in Figs. 5.15 and 5.17, the maximum solid-phase lithium-ion concentrations in the positive and negative electrodes of the simulation model are the identification results of the CCPSO algorithm and reference values, respectively.

Figure 5.17 displays the actual terminal voltage ($U1$) and simulated terminal voltage ($U2$) of ESP model or ESP-MCM. In Fig. 5.17, a, b or c, d offer a comparison between the new cell and pack and the degraded cell and pack, demonstrating a significant decrease in voltage after the same time of discharge. This decrease suggests a reduction in capacity performance due to battery degradation. Table 5.10 demonstrates the MAEs of terminal voltages in one cell of *Pack*2 and *Pack*2.

Figure 5.17 and Table 5.10 demonstrate that although the terminal voltage fitting effect of the ESP model and ESP-MCM under variable-rate energy storage conditions is not as good as that under low-rate constant current energy storage conditions, the simulation error is also small. Specifically, the MAEs of terminal voltage in the first and last cycles are compared in Table 5.10, revealing that the accuracy of ESP model and ESP-MCM slightly decreases with battery degradation, but remains within the

Table 5.9 The MAEs of SP and ESP model terminal voltages

Cell	One cell in *Pack*2 (SP model)	One cell in *Pack*2 (ESP model)
MAE	0.0113 V	0.0084 V

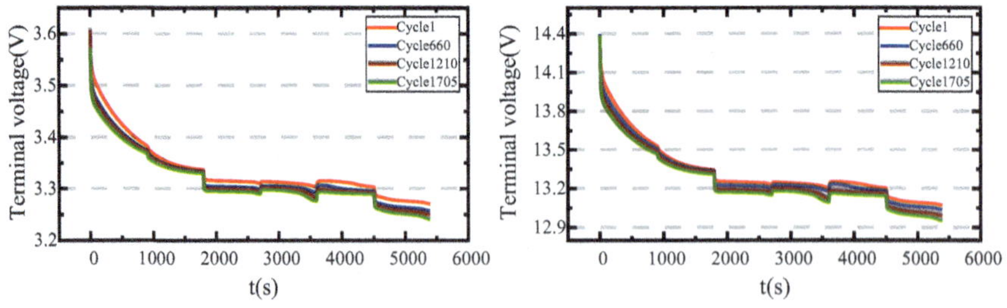

(a) The terminal voltages of one cell in *Pack* 2 (b) The terminal voltages of *Pack* 2

Fig. 5.16 The measured terminal voltages of *Pack*2

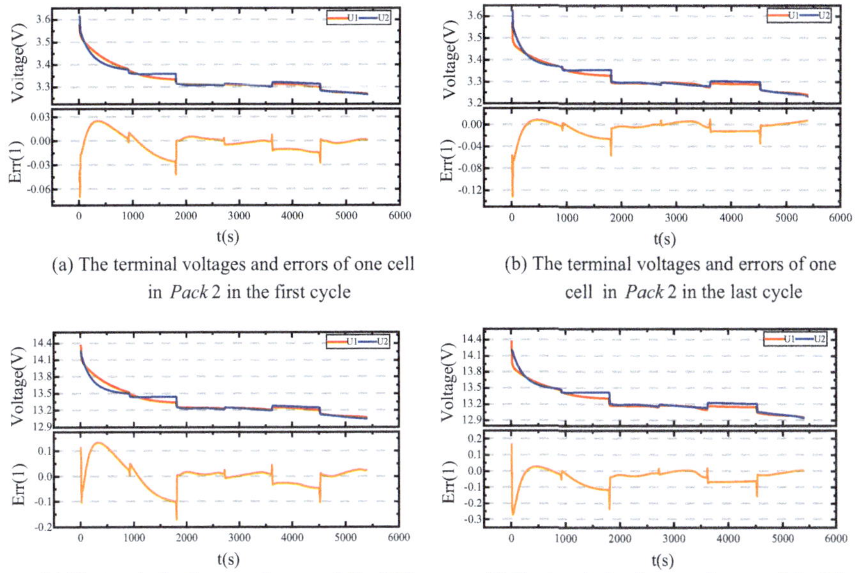

(a) The terminal voltages and errors of one cell in *Pack* 2 in the first cycle

(b) The terminal voltages and errors of one cell in *Pack* 2 in the last cycle

(c) The terminal voltages and errors of *Pack* 2 in the first cycle

(d) The terminal voltages and errors of *Pack* 2 in the last cycle

Fig. 5.17 Model verification curves of *Pack*2

Table 5.10 The MAEs of terminal voltages in one cell of *Pack*2 and *Pack*2

Cell/pack (voltage model)	MAE of the first cycle (V)	MAE of the last cycle (V)
One cell in *Pack*2 (ESP model)	0.0082	0.0085
*Pack*2 (ESP-MCM)	0.0364	0.0389

acceptable range. These findings confirm the effectiveness and superiority of the developed ESP model and ESP-MCM under variable-rate constant current energy storage degradation conditions.

5.4.2 Health State Estimation Method Validation Under Variable-Rate Constant Current Energy Storage Degradation Condition

Due to the fact that the quantification method of degradation modes can only be applied under low-rate constant current conditions, this section only uses the maximum solid-phase lithium-ion concentrations of positive and negative electrodes as the input of the neural network for training and testing the health state estimation model. Under the variable-rate energy storage degradation condition, it is also concluded that the test results of BP network and SA-BP network are poor. Therefore, this section also uses the NSA-BP neural network to improve the health state neural network estimation model. When training the model, the initial parameter settings in the neural network model are consistent with Table 5.7. When using data from *Cell2* for training, the training process of different neural network models is shown in Fig. 5.18.

In Fig. 5.18, E1, E2, and E3 represent BP neural networks based on electrochemical parameters, SA-BP neural networks based on electrochemical parameters, and NSA-BP neural networks based on electrochemical parameters, respectively. From Fig. 5.18, it can be seen that different neural network models need to train each individual number of times to meet the set target MAE requirements. The network connection weights and thresholds are recorded and used as model parameters for the test set. In this section, the relevant data of *Pack*2 is regarded as the test set. The diagrams of health state output from the test set and its error are shown in Fig. 5.19.

Figure 5.19 depicts the actual health state ($S1$), the health state based on BP network and $c_{s,\max,i}$ ($S2$), the health state based on simulated annealing-back propagation (SA-BP) network and $c_{s,\max,i}$ ($S3$), and the health state based on NSA-BP network and $c_{s,\max,i}$ ($S4$). $Err1$, $Err2$, and $Err3$ represent the same meaning corresponding to $S2$, $S3$, and $S4$, respectively. From Fig. 5.19, it can be seen that there is a small difference in the health state between the cells in *Pack*2, which also results in the overall health state curve of *Pack*2 not being completely consistent with the health state curve of each cell. The performance evaluation indicators for algorithms in *Pack*1 are presented in Table 5.11.

From Table 5.11, it can be seen that the health state estimation algorithm based on multidimensional HIs and NSA-BP model is obviously better than the previous algorithm in terms of RMSE and MAE error indicators, both for each cell in *Pack*2 and for the entire *Pack*2.

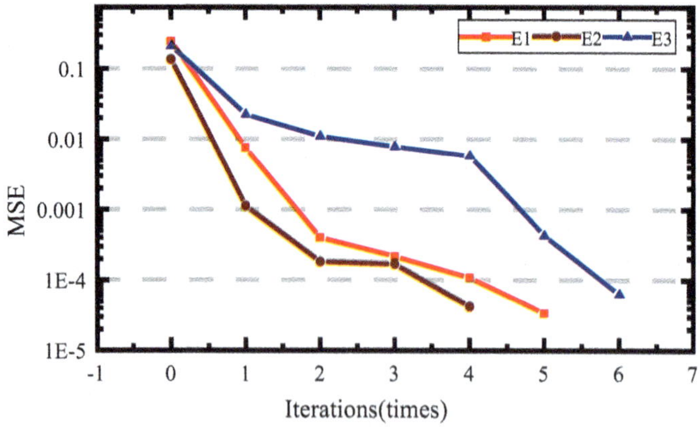

Fig. 5.18 The training process of three different neural network models

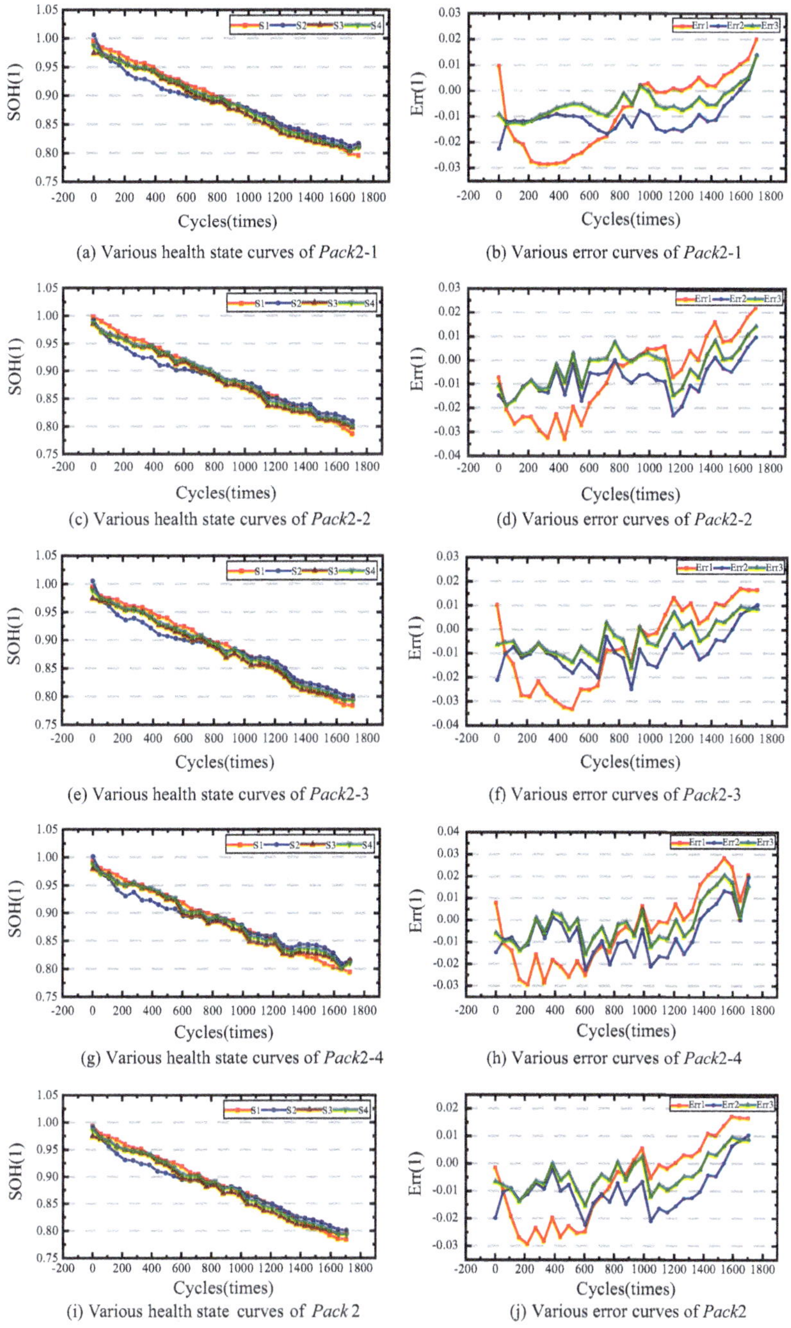

Fig. 5.19 The diagrams of health state output from *Pack*2 and its error

Table 5.11 The performance evaluation indicators for algorithms in *Pack*2

Type	*Pack*2-1	*Pack*2-2	*Pack*2-3	*Pack*2-4	*Pack*2
*S*2	RMSE: 0.0159 MAE: 0.0127	RMSE: 0.0167 MAE: 0.0136	RMSE: 0.0176 MAE: 0.0150	RMSE: 0.0170 MAE: 0.0144	RMSE: 0.0157 MAE: 0.0127
*S*3	RMSE: 0.0118 MAE: 0.0111	RMSE: 0.0104 MAE: 0.0087	RMSE: 0.0116 MAE: 0.0103	RMSE: 0.0119 MAE: 0.0103	RMSE: 0.0119 MAE: 0.0108
*S*4	RMSE: 0.0073 MAE: 0.0064	RMSE: 0.0085 MAE: 0.0067	RMSE: 0.0076 MAE: 0.0066	RMSE: 0.0095 MAE: 0.0078	RMSE: 0.0073 MAE: 0.0064

5.5 Summary

This chapter presents an experimental analysis of estimation methods aimed at verifying the universality of the health state estimation method. To accomplish this, features are extracted from actual energy storage condition curves, and different types of energy storage conditions are clustered using the K-means algorithm. Additionally, an experimental platform is established to conduct experiments on lithium-ion battery cells and battery packs under various typical energy storage conditions. The algorithm is then validated by exciting different lithium iron phosphate batteries under typical working conditions of energy storage. By comparing and analyzing the actual and estimated results of health state, the effectiveness of the novel estimation method proposed in this book is demonstrated.

References

1. Berecibar, M., et al. 2016. State of health estimation algorithm of $LiFePO_4$ battery packs based on differential voltage curves for battery management system application. *Energy* 103 (15): 784–796.
2. Safari, M., and C. Delacourt. 2011. Modeling of a commercial graphite/$LiFePO_4$ cell. *Journal of the Electrochemical Society* 158 (5): 562–571.